Suborbital

Industry at the Edge of Space

Other Springer-Praxis books of related interest by Erik Seedhouse

Tourists in Space: A Practical Guide
2008
ISBN: 978-0-387-74643-2

Lunar Outpost: The Challenges of Establishing a Human Settlement on the Moon
2008
ISBN: 978-0-387-09746-6

Martian Outpost: The Challenges of Establishing a Human Settlement on Mars
2009
ISBN: 978-0-387-98190-1

The New Space Race: China vs. the United States
2009
ISBN: 978-1-4419-0879-7

Prepare for Launch: The Astronaut Training Process
2010
ISBN: 978-1-4419-1349-4

Ocean Outpost: The Future of Humans Living Underwater
2010
ISBN: 978-1-4419-6356-7

Trailblazing Medicine: Sustaining Explorers During Interplanetary Missions
2011
ISBN: 978-1-4419-7828-8

Interplanetary Outpost: The Human and Technological Challenges of Exploring the Outer Planets
2012
ISBN: 978-1-4419-9747-0

Astronauts for Hire: The Emergence of a Commercial Astronaut Corps
2012
ISBN: 978-1-4614-0519-1

Pulling G: Human Responses to High and Low Gravity
2013
ISBN: 978-1-4614-3029-2

SpaceX: Making Commercial Spaceflight a Reality
2013
ISBN: 978-1-4614-5513-4

Erik Seedhouse

Suborbital

Industry at the Edge of Space

 Springer

Published in association with
Praxis Publishing
Chichester, UK

Dr Erik Seedhouse, M.Med.Sc., Ph.D., FBIS
Milton
Ontario
Canada

SPRINGER-PRAXIS BOOKS IN SPACE EXPLORATION

ISBN 978-3-319-03484-3   ISBN 978-3-319-03485-0 (eBook)
DOI 10.1007/978-3-319-03485-0
Springer Cham Heidelberg New York Dordrecht London

Library of Congress Control Number: 2013956603

Cover design: Jim Wilkie
Project copy editor: Christine Cressy

Printed on acid-free paper

Springer is part of Springer Science+Business Media (www.springer.com)

Contents

Acknowledgment.. ix
About the author... xiii
Figures.. xv
Tables... xix
Acronyms... xxi
Foreword.. xxv

1 Suborbital Spaceflight 1
Suborbital flight: a brief history.. 5
   Project Mercury 5
   X-15 11
The legacy of SpaceShipOne 15
Regulating the industry 19
Suborbital risks 20
   Decompression.. 20
   Acceleration 23
   Bail-out 28
   Vehicle design 30
   Emergency egress options... 30
   Radiation 30

2 Suborbital Market 33
The 10-year forecast 37
Reusable suborbital launch vehicles 38
Market analysis 40
The Markets 42
   Commercial human spaceflight... 42
   Basic and applied research.. 46
   Aerospace technology test and demonstration............................ 48
   Media and public relations.. 49

The *Apollo 13* space scenes 51
Education 53
Satellite deployment.. 54
Remote sensing 55
Point-to-point transportation.. 55

3   Training Suborbital Astronauts... 59
Medical standards 60
Suborbital medical environment 60
  Acceleration 62
  Microgravity effects 67
  Cardiovascular effects 67
  Neurovestibular effects 67
  X-15 neurovestibular experience 68
  Space motion sickness 69
  Emergency egress capability... 70
  Environmental medical issues... 71
  Radiation 72
  Noise 74
  Vibration 74
  Suborbital medical standards 74
Training for commercial suborbital spaceflight 76
  NASTAR 77
  Astronauts for Hire 79
  Suborbital Training 81
  Inner Space Training 81
  SIRIUS Astronaut Training 83

4   The Frontrunners.. 85
Virgin Galactic 85
  SS2 powered test flight 85
XCOR 92
  Lynx step by step 95

5   Contenders: Vehicles Waiting in the Wings.......................... 97
Blue Origin.. 98
Armadillo Aerospace 102
Masten Space Systems 104
Copenhagen Suborbitals 107

6   Spaceports.. 113
Spaceport America.. 116
Caribbean Spaceport 117
Spaceport Sweden 118
Mojave Spaceport 123

7 Suborbital Science... 125
Potential suborbital science potential capabilities 127
Portfolio of game-changing missions 127
  The stand test 130
Anatomy of a science mission 131
Making the most of those four minutes 136
  Eyes versus ears: A primer on motion sickness............................... 137
Launch day... 138

8 Payloads 141
Flying a payload with virgin galactic.. 141
  SS2 Timeline... 142
  Microgravity environment and G-loading 144
Flying a payload with XCOR 148
Flying a payload with Blue Origin.. 152
The suborbital payload agent 153

9 How to Get There.. 155
Strategies to fly a mission 158
Become a citizen-astronaut 158
Fly a payload.. 160
Win a ticket 161
Perform research 162

Appendix I 165
Appendix II.. 169
Appendix III 175
Appendix IV 177

Index.. 181

Acknowledgments

In writing this book, the author has been fortunate to have had five reviewers who made such positive comments concerning the content of this publication. He is also grateful to Maury Solomon at Springer and to Clive Horwood and his team at Praxis for guiding this book through the publication process. The author also gratefully acknowledges all those who gave permission to use many of the images in this book.

The author also expresses his deep appreciation to Christine Cressy, whose attention to detail and patience greatly facilitated the publication of this book, to Rekha Udaiyar for her meticulous attention to detail in proofing the book, and to Jim Wilkie for creating the cover.

From L to R: MiniMach, Lava, and Jasper

To our cats, Jasper, Mini-Mach, and Lava – an
integral part of providing comfort in our lives

About the author

Erik Seedhouse is a Norwegian-Canadian suborbital astronaut whose life-long ambition is to work in space. After completing his first degree in Sports Science at Northumbria University, the author joined the legendary 2nd Battalion the Parachute Regiment, the world's most elite airborne regiment. During his time in the "Para's", Erik spent six months in Belize, where he was trained in the art of jungle warfare. Later, he spent several months learning the intricacies of desert warfare in Cyprus. He made more than 30 jumps from a Hercules C130, performed more than 200 helicopter abseils, and fired more anti-tank weapons than he cares to remember!

Upon returning to the comparatively mundane world of academia, the author embarked upon a master's degree in Medical Science at Sheffield University. He supported his studies by winning prize money in 100-km running races. After placing third in the World 100 km Championships in 1992 and setting the North American 100-km record, the author turned to ultra-distance triathlon, winning the World Endurance Triathlon Championships in 1995 and 1996. For good measure, he also won the inaugural World Double Ironman Championships in 1995 and the Decatriathlon, a diabolical event requiring competitors to swim 38 km, cycle 1,800 km, and run 422 km. Non-stop!

Returning to academia in 1996, Erik pursued his Ph.D. at the German Space Agency's Institute for Space Medicine. While conducting his studies, he found time to win Ultraman Hawaii and the European Ultraman Championships as well as completing Race Across America. Due to his success as the world's leading ultra-distance triathlete, Erik was featured in dozens of magazines and television interviews. In 1997, *GQ* magazine nominated him as the "Fittest Man in the World".

In 1999, Erik retired from being a professional triathlete and started post-doctoral studies at Simon Fraser University. In 2005, he worked as an astronaut training consultant for Bigelow Aerospace and wrote *Tourists in Space*, a manual for spaceflight participants. He is a Fellow of the British Interplanetary Society and a member of the Space Medical Association. In 2009, he was one of the final 30 candidates in the Canadian Space Agency's Astronaut Recruitment Campaign. Erik works as a corporate astronaut (www.suborbital-training.com), spaceflight consultant, professional speaker, triathlon coach, and author.

He is the Training Director for Astronauts for Hire (www.astronauts4hire.org) and completed his suborbital astronaut training in May 2011. Between 2008 and 2012, he served as director of Canada's manned centrifuge operations.

In addition to being a suborbital astronaut, triathlete, centrifuge operator, pilot, and author, Erik is an avid mountaineer and is pursuing his goal of climbing the Seven Summits. *Suborbital* is his 12th book. When not writing, he spends as much time as possible in Kona on the Big Island of Hawaii and at his real home in Sandefjord, Norway. Erik and his wife, Doina, are owned by three rambunctious cats – Jasper, Mini-Mach, and Lava.

Figures

1.1	Astronaut Mike Melvill	2
1.2	XCOR's Lynx spacecraft with dorsal payload	3
1.3	SpaceShipTwo undergoing glide tests	3
1.4	Neil Armstrong in the cockpit of the X-15	4
1.5	Ham following his epic flight	6
1.6	Al Shepard	7
1.7	Al Shepard's Mercury launch	8
1.8	Gus Grissom outside the Liberty Bell	10
1.9	X-15	12
1.10	X-15 slung under its mother ship	13
1.11	X-15 landing, just like SpaceShipOne	14
1.12	WhiteKnight1 and SpaceShipOne	15
1.13	SpaceShipOne interior	16
1.14	Brian Binnie flies the flag after SpaceShipOne wins the X-Prize	17
1.15	The aftermath of a rapid decompression	21
1.16	(a) Pressure suit; (b) Nic of Final Frontier Design	22
1.17	Signs and symptoms of altitude decompression sickness	23
1.18	The G-LOC syndrome	24
1.19	The author seated in Canada's only centrifuge	26
1.20	Felix Baumgartner waits for the moment of truth	29
1.21	Sea survival training	31
2.1	Suborbital reusable launch vehicle	34
2.2	Potential growth in the suborbital markets	36
2.3	Potential flight revenue in the suborbital markets	37
2.4	Tauri Group's research strategy	38
2.5	The NanoRack	39
2.6	SpaceShipTwo glides to a landing in the Mojave	40
2.7	Reaction Rocket Engines' revolutionary transportation system	42
2.8	Richard Garriott	43

2.9   Black Brant sounding rocket 50
2.10   Reaction Rocket Engines' SABRE engine 57
2.11   Waverider 58

3.1   Human centrifuge at Wyle Labs 61
3.2   ESA uses a modified Airbus for its parabolic flights 61
3.3   The different types of acceleration 62
3.4   Mercury-Redstone 3 launch 64
3.5   The Johnsville centrifuge 65
3.6   The author being fitted for his G-suit in preparation for his Hawk flight 66
3.7   Dunker training 70
3.8   The author prepares for a high-altitude indoctrination chamber flight 72
3.9   A solar flare 73
3.10   The NASTAR centrifuge 78
3.11   Members of Astronauts for Hire 79
3.12   Unusual attitude flying as performed by a pair of NASA T-38s 80

4.1   Experiencing weightlessness on board SpaceShipTwo 86
4.2   Spaceport America 86
4.3   Virgin Galactic logo, featuring Richard Branson's eye 88
4.4   Technical specifications of SpaceShipTwo 90
4.5   The view from space 91
4.6   The author in the passenger seat of the Lynx mock-up 93
4.7   Spaceport Curaçao 93

5.0   Copenhagen Suborbitals' mission patch for the Tycho Brahe 97
5.1   Blue Origin's New Shepard spacecraft 98
5.2   Delta Clipper 100
5.3   Armadillo Aerospace 102
5.4   Armadillo's STIG-B 103
5.5   Xombie 105
5.6   Copenhagen Suborbitals' Tycho Brahe 108
5.7   The Horizontal Assembly Building (HAB) 108
5.8   Test flight of the Tycho Brahe 110

6.1   Dunker training 114
6.2   Spaceport America 116
6.3   Aurora borealis 119
6.4   Rocket launch from Kiruna's Esrange 120
6.5   Inside Kiruna's IceHotel 121
6.6   Mojave Air and Space Port 122

7.1   Alan Stern after his ride in the Starfighter 126
7.2   Virgin Galactic payload configuration 128
7.3   Noctilucent clouds 128

7.4   Terrestrial gamma-ray flashes 129
7.5   Jason Reimuller, Principal Investigator of the PoSSUM flight 131
7.6   Polar mesospheric clouds 132
7.7   PoSSUM patch 133

8.1   Lynx Payload configuration 142
8.2   Payload User Guide 143
8.3   SpaceShipTwo flight profile 144
8.4   Sleeve bolt receptacles on SpaceShipTwo 146
8.5   Lynx payload configuration 147

9.1   SpaceShipTwo slung under its mother ship, WhiteKnightTwo 157
9.2   XCOR's Cub Carrier 159
9.3   Water bears are a candidate for suborbital life sciences flights 160
9.4   (a) Tale Sundlisaeter in the back seat of an F-16; (b) Tale as she looks
        without fighter pilot gear 163

Tables

1.1	Mercury-Redstone Mission Sequence	9
1.2	Manned Suborbital Flights	18
2.1	SRLVs in Development	34
2.2	SRLV Markets	35
2.3	Seat/Cargo Equivalents	35
2.4	Forecasts for the Commercial Human Spaceflight Market	46
2.5	Technology Readiness Levels	49
2.6	Satellite Categories	54
2.7	Remote Sensing Market	56
3.1	Spaceflight Participant Questionnaire	76
3.2	Suborbital Training	82
3.3	Topics Included in Inner Space Training	83
5.1	Tycho Brahe	110
6.1	Spaceport Features	115
7.1	Timeline for a Suborbital Science Flight	134
7.2	Suggested Events Leading up to Suborbital Science Launch	136
7.3	Launch Countdown Milestones	139
8.1	Expected G-Loads for Flight and Crash Conditions Direction	144
8.2	Payload Specifications Type	147
8.3	Lynx Variants Capabilities	148
8.4	XCOR Payload Locations	150
8.5	Lynx Preflight Timeline	152
8.6	Cabin Payload Bays	153

Acronyms

A4H	Astronauts for Hire
ADS	Air Data System
AGSM	Anti-G Straining Maneuver
AGSOL	Ashton Graybiel Spatial Orientation Laboratory
ALASA	Airborne Launch Assist Space Access
ALOC	Almost Loss of Consciousness
BWE	Brainwave Entrainment
CCDev	Commercial Crew Development
CCP	Commercial Crew Program
CFD	Computational Fluid Dynamics
CS	Copenhagen Suborbitals
CSA	Canadian Space Agency
CSF	Commercial Space Foundation
DCS	Decompression sickness
DC-X	Delta Clipper Experimental
EEG	Electroencephalogram
EELV	Evolved Expendable Launch Vehicle
EMI	Electromagnetic Interference
EPT	Effective Performance Time
ESA	European Space Agency
ETC	Environmental Tectonics Corporation
FAA	Federal Aviation Administration
FAI	Fédération Aéronautique Internationale
FFD	Final Frontier Design
FTE	Flight Test Engineer
G-LOC	Gravity-Induced Loss of Consciousness
GN&C	Guidance Navigation & control
GOR	Gradual Onset Run
GPS	Global Positioning System
GSA	General Sales Agent

HAB	Horizontal Assembly Building
HATV	Hybrid Atmospheric Test Vehicle
HTP	High Test Peroxide
ICRP	International Commission on Radiological Protection
INS	Inertial Navigation System
IR	Infrared
ISS	International Space Station
ITAR	International Trade in Arms Regulations
JPL	Jet Propulsion Laboratory
LEO	Low Earth Orbit
LOX	Liquid Oxygen
MART	Multi-Axis Rotation and Tilt
MDF	Mild Detonating Fuse
NACA	National Advisory Committee for Aeronautics
NASTAR	National Aerospace Training and Research
NCRP	National Council for Radiation Protection
NMSA	New Mexico Spaceport Authority
NRO	National Reconnaissance Organization
NSBRI	National Space Biomedical Research Institute
NSRC	Next Generation Suborbital Researchers Conference
NTPS	National Test Pilot School
P2P	Point-to-Point
PI	Principal Investigator
PLL	Peripheral Light Loss
PMC	Polar Mesospheric Cloud
PoSSUM	Polar Suborbital Science in the Upper Mesosphere
PSI	Planetary Science Institute
PUG	Payload User Guide
RCS	Reaction Control System
REM	Research Education Mission
RLV	Reusable Launch Vehicle
ROR	Rapid Onset Run
ROSES	Research Opportunities in Earth and Space Sciences
SARG	Suborbital Applications Research Group
SAS	Space Adaptation Syndrome
SCR	Solar Cosmic Radiation
SFP	Spaceflight Participant
SMS	Space Motion Sickness
SPA	Shuttle Payload Adapter
sRLV	Suborbital Reusable Launch Vehicle
SS1	SpaceShipOne
SS2	SpaceShipTwo
SSTO	Single Stage to Orbit
SWORDS	Soldier-Warfighter Operationally Responsive Deployer for Space
SwRI	Southwest Research Institute

SXC	Space Expedition Corporation
TGF	Terrestrial Gamma-ray Flashes
TPS	Thermal Protection System
TRL	Technology Readiness Level
TUC	Time of Useful Consciousness
TVC	Thrust Vector Control
UAV	Uninhabited Aerial Vehicle
USAF	United States Air Force
USML	United States Munitions List
USN	United States Navy
USRA	Universities Space Research Association
UV	Ultraviolet
VTOL	Vertical Take-Off and Landing
WK1	WhiteKnight1
WK2	WhiteKnight2

Foreword

"We're really on the cusp of an exciting new capability for our country and for our economy."

Lori Garver, NASA's deputy administrator, explaining why NASA is seeking US$75 million for NASA's Commercial Reusable Suborbital Research program

As the main engine ignites, the crew feels a deep rumble far below them and a sudden sensation of motion as the launch vehicle lifts off, trailing a 150-meter-long fountain of sun-bright exhaust in an inferno of smoke, searing light, and earth-shaking noise. The mix of payload specialists and scientists feel the thunder of the launch, the numbing noise, and the incredible acceleration, as they are pushed forcefully back into their seats. The gut-wrenching journey to suborbit, an event planned for many weeks and anticipated by the crew for several months, takes less than five minutes. Once in microgravity, the thrill of the ascent is replaced by the immediacy of the moment, as the crew pull out checklists and go to work on their payloads and experiments, mindful that they have less than five minutes to complete their task-lists. Welcome to the world of suborbital spaceflight.

Until recently, spaceflight had been the providence of a select corps of professional astronauts whose missions, in common with all remarkable exploits, were experienced vicariously by the rest of the world via television reports and internet feeds. These space-farers risked their lives in the name of science, exploration, and adventure, thanks to government-funded manned spaceflight programs.

All that is about to change. In fact, it is changing as you read these very words.

The nascent commercial suborbital spaceflight industry will soon open the space fron-tier to commercial astronauts, payload specialists, scientists, and, of course, spaceflight participants. *Suborbital* describes the tantalizing science opportunities offered when sub-orbital trips become routine in the 2014 to 2016 timeframe. It describes the difference in training and qualification necessary to become either a spaceflight participant or a fully fledged commercial suborbital astronaut and it describes the vehicles this new class of astronauts will use.

Anticipation is on the rise for the new crop of commercial suborbital spaceships that will serve the scientific and educational market. These reusable rocket-propelled vehicles are expected to offer quick, routine, and affordable access to the edge of space, along with the capability of carrying research and educational crewmembers. Yet to be demonstrated is the hoped-for flight rates of suborbital vehicles. Quick turnaround of these craft is central to realizing the profit-making potential of repeated sojourns to suborbital heights. As this proposal outlines, vehicle builders still face rigorous shake-out schedules, flight-safety hurdles, as well as extensive trial runs of their respective craft before suborbital space jaunts become commonplace. *Suborbital* examines some of these "cash and carry" suborbital craft under development by such groups as Blue Origin, Masten Space Systems, Virgin Galactic, and XCOR Aerospace, and describes the hurdles the space industry is quickly overcoming en route to the industry developing into a profitable economic entity.

Suborbital also explains how the commercial suborbital spaceflight industry is planning and preparing for the challenges of marketing and financing and how it is marketing the hiring of astronauts. It examines the role of commercial operators as enablers accessing the suborbital frontier and how a partnership with governments and the private sector will eventually permanently integrate the free market's innovation of commercial suborbital space activities.

1

Suborbital Spaceflight

For those of you unfamiliar with the suborbital flight industry, this chapter provides you with a primer. And, if you're one of the lucky few who have the financial means to buy a ticket, this chapter gives you an idea of what you should know before you pull on that spacesuit.

Those of you who have followed the suborbital spaceflight industry over the last few years have cause to be disappointed because it's been quite a waiting game. After the euphoria of the X-Prize in 2004, space fans talked about the possibility of flying into space the following year or, if not the following year, then *definitely* the year after that. The operators and the companies that comprised the nascent commercial spaceflight industry fueled the speculation that suborbital revenue flights were imminent by making promises they would soon be ready to fly to space with you and your friends on board. One year stretched to two and two became five. Now, in 2014, almost a decade after SpaceShipOne's (SS1) historic flight (Figure 1.1), it looks like the industry is finally tantalizingly within reach of realizing the promise of the X-Prize.

That message was already evident in the presentations made by the five major suborbital vehicle developers – Armadillo Aerospace, Blue Origin, Masten Space Systems, Virgin Galactic, and XCOR Aerospace – at the Next-Generation Suborbital Researchers Conference (NSRC) in Palo Alto, California, in February 2012. The conference, the third of its kind, attracted more than 400 people interested in performing scientific research and technology demonstrations on the vehicles these companies are developing. Among the developments announced at the meeting was the US$5 million round of funding XCOR Aerospace had secured to complete the Mark I prototype of its Lynx suborbital vehicle (Figure 1.2).

Since NSRC 2012, XCOR has been making steady progress on its Lynx, a winged vehicle that will take off from a runway under its own rocket power and ascend to altitudes of about 60 kilometers (the Mark II will reach 100 kilometers or higher) before gliding back to a runway. Meanwhile, other companies have also been busy building and flying their vehicles. The leader of the pack, Virgin Galactic, announced in December 2012 that its SpaceShipTwo (SS2) had completed its first glide in powered flight configuration (Figure 1.3), the vehicle's 23rd glide flight in the pre-powered portion of its incremental test flight program.

E. Seedhouse, *Suborbital: Industry at the Edge of Space*, Springer Praxis Books,
DOI 10.1007/978-3-319-03485-0_1, © Springer International Publishing Switzerland 2014

1.1 Astronaut Mike Melvill after his spaceflight on September 29th, 2004. Courtesy: Scaled Composites

The flight was significant, as it was the first test with rocket motor components installed, including tanks, and it was also the first flight with thermal protection applied to the vehicle's leading edges. Since NSRC 2012, the secret squirrel suborbital enterprise known as Blue Origin, the company created by Amazon.com founder, Jeff Bezos, had also been making headlines with the successful test of its crew escape system in October 2012. The test included a firing of the escape motor on a full-scale suborbital crew capsule that reached an altitude of 703 meters. The crew escape system is designed for the company's suborbital vehicle and uses a pusher motor rather than the tractor system that "pulled" the capsule away from rockets during the early days of the manned space program and which is still used on the Soyuz.

The successes of the suborbital companies before and since NSRC 2012 echoed the endorsement of a former X-15 test pilot better known for his time as a NASA astronaut: Neil Armstrong (Figure 1.4). Armstrong's keynote at NSRC 2012 focused mainly on the X-15 which, he noted, was built for speed, not altitude, although the vehicle did go on to achieve suborbital altitudes. He didn't directly address the commercial suborbital vehicle operators in his talk, but he expressed hope they would find success.

1.2   XCOR's Lynx spacecraft with dorsal payload. Courtesy: XCOR

1.3   SpaceShipTwo undergoing glide tests. Courtesy: Mark Greenberg/Virgin Galactic

1.4 Neil Armstrong in the cockpit of the X-15. Courtesy: NASA

"I think that this shows that commercial suborbital, while being leading edge, is also becoming mainstream. I think Neil's being here is a pretty certain demonstration of that. The fact that he wanted to travel across the country – he actually canceled an overseas trip so he could speak at a suborbital conference – I think speaks volumes."

Alan Stern, associate vice president of the Southwest Research Institute and NSRC organizer

That said, operators still need to deliver on their promises because, in the world of manned spaceflight, whether it be orbital or suborbital, schedules slip. A case in point: at the first NSRC in 2010, Virgin Galactic and XCOR said they would be conducting powered flight tests of their vehicles in 2011. It didn't happen. And, at NSRC 2012, companies still talked about potential delays. At NSRC 2013, the story was the same. But, with so many companies in the game, each with differing technical approaches, the era of routine, reliable, low-cost suborbital space access may finally soon be realized. I certainly hope so, because I'm one of the many who want to fly! But, before describing how you and I can climb on board SS2 or the Lynx, it's worth taking a moment to understand what exactly what suborbital flight is.

SUBORBITAL FLIGHT: A BRIEF HISTORY

Project Mercury

In short, a suborbital flight is defined as a flight to an altitude higher than 100 kilometers that does not involve sending a vehicle into orbit. The first unmanned suborbital flight was in early 1944, when a V-2 test rocket launched from Peenemünde, in Germany, reached an altitude of 189 kilometers. Then, on February 24th, 1949, the upper stage of Bumper 5, a two-stage rocket launched from the White Sands Proving Grounds, reached an altitude of 399 kilometers. These flights were followed by the first pioneers of the space age: monkeys (see Appendix I). You may be wondering what a monkey has to do with the space program but to answer that question we have to go back to the 1950s. As part of the space race with the Soviet Union, Project Mercury (1958–63) was tasked with putting an American astronaut into orbit and returning him safely. The program was also designed to test how well humans functioned in the space environment. But before humans could be launched, NASA needed to make sure their astronauts could be kept safe from micrometeoroids, radiation, noise, vibration, G-forces, microgravity, and the vacuum of space. Also, medical experts were unsure whether humans could handle being isolated and confined in the claustrophobic interior of a space capsule. So, before taking a risk with a human, scientists recruited *astro-chimps*. Chimpanzees were an obvious choice because a chimp's organ and skeletal structures are similar to ours, and chimps can be trained. First up was Gordo, a squirrel monkey launched by the US Army on board a Jupiter AM-13 booster on December 13th, 1958. Gordo made the suborbital flight with no adverse effects, but paid the ultimate price because the flotation mechanism of the rocket's nose cone failed. Next in line were Able and Baker. Able was a female rhesus monkey and Baker was a female squirrel monkey. Instrumented with electrodes to monitor their vital signs during the flight, the intrepid pair was launched on May 28th, 1959, on another Jupiter booster. Both survived the 480-kilometer flight and were recovered. Sadly, Able died on an operating table as doctors performed surgery to remove the sensors from underneath her skin. Six months later, Sam – a rhesus monkey – was launched on board the first US Little Joe flight on December 4th, 1959. He survived the flight and was recovered. Then there was Chop Chop Chang, who was launched on board a Mercury-Redstone 2 rocket on January 31st, 1961. Constructed of titanium just 0.25 millimeters thick, blanketed by fiberglass insulation and covered with blackened heat-radiating shingles, the rocket's sole passenger was the four-year-old ape. Chop Chop Chang, or Number 65 (Figure 1.5) as he was also referred to, was, by all accounts, a smart, loveable chimp with a positive temperament – ideal astronaut material in other words. Weighing 16 kilograms, Number 65's mission was to test the environmental control systems inside the Mercury capsule and to determine whether the Mercury-Redstone rocket was safe for humans and whether primates could function under the stress and pressure that come with space travel. Ensconced in his coffin-sized "cockpit", Number 65 was anything but a passenger because he had a number of tasks. For the correct response he earned a banana pellet and for a wrong response he received an electrical shock. There is no indication whether NASA tried this type of training with their astronauts!

Number 65's flight demonstrated that chimps could work in flight. Through the launch, more than six minutes of weightlessness, and re-entry, Number 65 moved levers in response to flashing lights, just as he had been taught in the lab. In fact, his response times in space

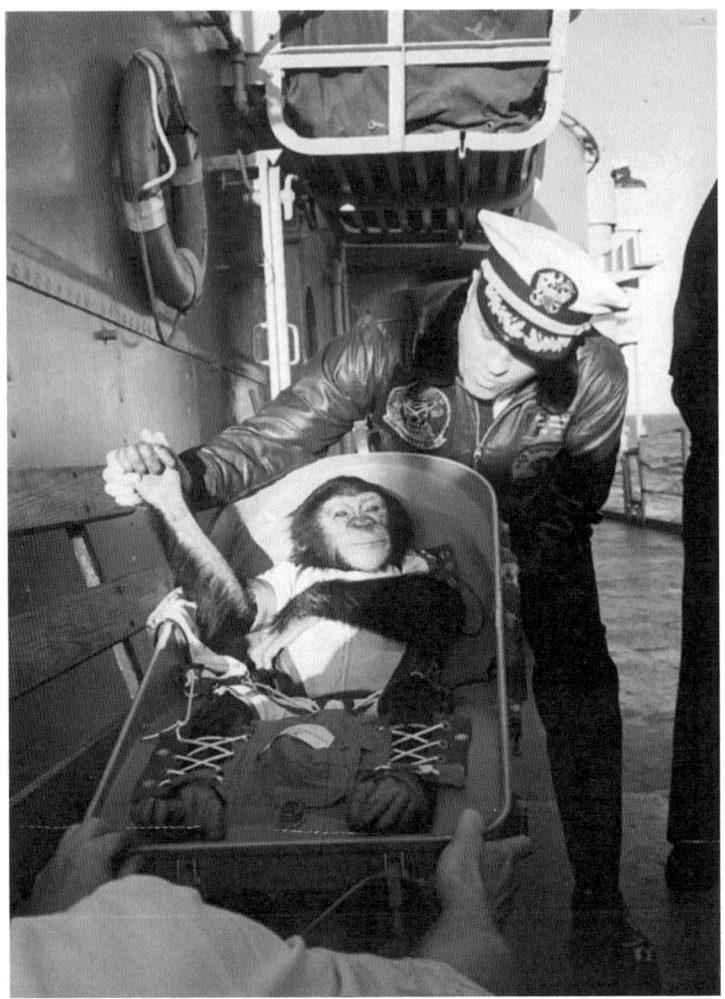

1.5   Ham following his epic flight. Courtesy: US Navy/NASA

were as good as on Earth. Unfortunately, the fuel in his rocket burned off too quickly and he was propelled almost 200 kilometers further than planned. He also experienced acceleration forces up to 14.7 G, which was 3.3 G more than planned. Making matters worse, Number 65's capsule made a rough landing downstream, the impact upon hitting the ocean causing his capsule to take on water. Waiting to recover the capsule were six naval destroyers and a landing ship dock, the USS *Donner*, with three helicopters. Unfortunately, because of the launch glitch, the task force was waiting in the wrong place. In case of such an event, four surveillance aircraft were ready to search for the capsule and, half an hour after landing, a plane spotted the capsule and helicopters were sent to collect it. Back on the ship, Number 65 was finally taken out of the capsule. He shook hands with the captain, ate an apple and half an orange, and was checked by a doctor who pronounced the chimp to be in good condition, despite the fact his 16-minute flight had extended to a rescue mission

1.6 Al Shepard. Courtesy: NASA

lasting almost four hours. In fact, Number 65 showed no ill effects from his flight. Well, almost no ill effects. Seems Number 65, whose flight had earned him a new name – Ham – derived from his home unit, the Holloman Aeromedical Laboratory, wasn't too taken with the whole spaceflight experience because when the press wanted photos of him in his couch he fought to avoid being strapped in. Nevertheless, the flight was pronounced a success and Ham (the main character of the 2008 movie *Space Chimps* was named Ham III, the grand-son of Ham) became a cause for celebration. He landed on the cover of *Life Magazine*, and was covered by all the newsreels. The Mercury astronauts were especially pleased Ham had suffered no ill effects because they knew it wouldn't be long before one of them would be strapped on top of the Mercury-Redstone rocket.

Thanks to the flights of Ham and co., NASA decided the Mercury-Redstone rocket was safe enough to launch an astronaut. So, in May 1961, Alan Shepard (Figure 1.6)

1.7   Al Shepard's Mercury launch. Courtesy: NASA

clambered on board a Mercury capsule on the man-rated version Mercury-Redstone 3 and launched (Figure 1.7) on a suborbital flight reaching an altitude of 187 kilometers (Table 1.1).

Table 1.1. Mercury-Redstone Mission Sequence.

Time (min/sec)	Event	Description
00:00	Lift-off	Mercury-Redstone lifts off
00:16	Pitch Program	Redstone pitches over 2°/sec from 90° to 45°
00:40	End Pitch Program	Redstone reaches 45° pitch
01:24	Max Q	Maximum dynamic pressure
02:20	BECO	Engine shutdown – Booster Engine Cut-Off Velocity 2.3 km/sec
02:22	Tower Jettison	Escape Tower Jettison
02:24	Spacecraft Separation	Posigrade rockets fire for 1 sec, giving 4.6 m/sec separation
02:35	Turnaround Maneuver	ASCS rotates spacecraft 180°, to heat shield forward attitude; nose pitched down 34° to retro fire position
05:00	Apogee	Apogee of 185 km reached 240 km downrange from launch site
05:15	Retrofire	Three retro rockets fire for 10 sec at 5-sec intervals; 170 m/sec is taken off forward velocity
05:45	Retract Periscope	Periscope is automatically retracted in preparation for re-entry
06:15	Retro Pack Jettison	Retro pack is jettisoned, leaving heat shield clear
06:20	Retro Attitude Maneuver	ASCS orients spacecraft 34° nose-down pitch, 0° roll, 0° yaw
07:15	0.05 G (0.5 m/sec²) Maneuver	ASCS detects beginning of re-entry and rolls spacecraft at 10°/sec to stabilize spacecraft during re-entry
09:38	Drogue Parachute Deploy	Drogue parachute deployed at 6.7 km slowing descent to 111 m/sec and stabilizing spacecraft
09:45	Snorkel Deploy	Fresh air snorkel deploys at 6.1 km; ECS switches to emergency oxygen rate to cool cabin
10:15	Parachute Deploy	Main parachute deploys at 3 km. Descent rate slows to 9.1 m/sec
10:20	Landing Bag Deploy	Landing bag deploys, dropping heat shield down 1.2 m
10:20	Fuel Dump	Remaining hydrogen peroxide fuel automatically dumped
15:30	Splashdown	Spacecraft lands in water 480 km downrange from launch site
15:30	Rescue Aids Deploy	Rescue aid package deployed

Shepard's flight was followed by Mercury-Redstone 4 (MR-4) two months later, piloted by Gus Grissom (Figure 1.8). Engineers had improved Grissom's spacecraft by adding a large viewing window and an explosively actuated side hatch. The hatch utilized an explosive charge to fracture the attaching bolts and thus separate the hatch from the spacecraft. Securing the hatch to the doorsill were titanium bolts and drilled in each bolt was a 1.5 millimeter-diameter hole which provided a weak point. A mild detonating fuse (MDF) was installed in a channel between an inner and outer seal around the periphery of the hatch. When the MDF was ignited, the resulting gas pressure between the inner and outer

1.8   Gus Grissom outside the Liberty Bell. Courtesy: NASA

seal would result in the bolts failing. The MDF was ignited by a manually operated igniter after the removal of a safety pin. If necessary, the igniter could be operated externally by an attached lanyard. This last item was to cause problems at the end of Grissom's mission.

On March 7th, 1961, the spacecraft was delivered to Hanger S at Cape Canaveral, where instrumentation and items of the communication system were removed for testing. After the items were reinstalled, the systems test proceeded as scheduled. The tests, which took 33 days, examined the electrical, sequential, instrumentation, communication, environmental, reaction-control, and stabilization/control systems. After these tests had been completed, the landing impact bag was installed and a simulated flight was run, after which the parachutes and pyrotechnics were installed and the spacecraft was delivered to the launch pad, where it spent another 21 days. Launch finally took place on July 21st, 1961, at 7:20 a.m. EST. The launch had originally been scheduled for July 18th, 1961, but was rescheduled to July 19th, due to unfavorable weather conditions. The launch attempt on July 19th was canceled at T−10 minutes due to more unfavorable weather. The first half of the split launch countdown began at 6:00 a.m. EST on July 20th, at T−640 minutes and proceeded normally through the 12-hour planned hold period for hydrogen peroxide and pyrotechnic servicing. The second half of the countdown proceeded at 2:30 a.m. EST on July 20th. At T−180 minutes, prior to liquid oxygen (LOX) loading, a planned 1-hour hold was called for another weather assessment. The assessment was favorable and the count proceeded at 3:00 a.m. EST. At T−45 minutes, a 30-minute hold was required to install a misaligned hatch bolt and, at T−30 minutes, a nine-minute hold was called to turn off the pad searchlights which interfered with launch telemetry. Then, at T−15 minutes, a 41-minute hold was called to wait for better cloud conditions, after which the count proceeded until lift-off. By the time launch finally rolled around, Grissom had spent 3 hours and 22 minutes in the spacecraft.

At T−35 seconds, the spacecraft umbilical was pulled and the periscope retracted. During the boost phase of flight, the flight-path angle was controlled by the launch-vehicle control system. Launch-vehicle cut-off occurred at T+2 minutes 23 seconds, at which time the escape tower was released. Ten seconds later, the spacecraft-to-launch-vehicle adapter clamp ring was separated, and the posigrade[1] rockets fired to separate the spacecraft from the launch vehicle. The periscope was extended, and the automatic stabilization and control system oriented the spacecraft to orbit attitude. Retro sequence was initiated at T+4 minutes 46 seconds, which was 30 seconds prior to the spacecraft reaching its apogee. Grissom assumed control of the spacecraft's attitude at T+3 minutes 5 seconds and controlled the spacecraft by the manual proportional control system to T+5 minutes 43 seconds. He initiated firing of the retrorockets at T+5 minutes 10 seconds and, from T+5 minutes 43 seconds, he controlled the spacecraft by the manual rate command system through re-entry. The retrorocket package jettisoned at T+6 minutes 7 seconds, the drogue chute deployed at T+9 minutes 41 seconds, and the main parachute at T+10 minutes 14 seconds. The flight was successful, but the spacecraft was lost during the post-landing recovery period as a result of premature actuation of the explosively actuated side egress hatch, resulting in the capsule sinking shortly after splashdown. Grissom exited the spacecraft immediately after hatch actuation and was quickly retrieved.

X-15

"This has been an area that has been essentially absent for about four decades, after the X-15 finished its job. There's a lot of work to be done there, and there's a lot of opportunity."

Spaceflight legend and historical icon, Neil Armstrong, speaking at the Next-Generation Suborbital Researcher's Conference, Palo Alto, February 2012

To the east of Mojave is Edwards Air Force Base, home for legendary pilots and the milestone-making NASA X-15 rocket plane (Figure 1.9), which roared to life high above the Mojave landscape for nearly a decade (the program included 199 flights between 1959 and 1968). Flying close to the edge of space, and sometimes beyond, the X-15 flights mimicked SS1's private trek to space more than three decades later.

---

[1] An auxiliary rocket on a multistage spacecraft that is fired in the direction of the spacecraft's motion to separate the sections.

1.9 X-15. Courtesy: NASA

The X-15 program was a joint effort of NASA, the US Air Force (USAF), and the US Navy (USN). During the program, eight pilots[2] met the US criterion for spaceflights by passing an altitude of 80 kilometers and were awarded astronaut status. Two of the X-15 pilots also qualified for recognition by the Fédération Aéronautique Internationale (FAI). After its initial test flights in 1959, the X-15 became the first winged aircraft to reach hypersonic velocities of Mach 4, 5, *and* 6, and to operate at altitudes above 30,500 meters. In common with SS21 and SS2, the X-15 was carried under the wing of another aircraft (Figure 1.10), in this case a Boeing B-52 bomber – Scaled Composites built their own B-52, which they called WhiteKnight.

The wedge-shaped tail surfaces of the X-15 provided directional stability at speeds where conventionally shaped airfoils would have had little effect. The large upper and lower fins and the downward slant of the wings allowed the X-15 to remain stable during steep climbs and at high altitudes. Covering the titanium substructure was a coating of Inconel X, a nickel alloy capable of withstanding temperatures of 650°C (more heat dissipation was achieved by coloring the fuselage black). Not surprisingly, Inconel X, like the airframe, thermal protection, flight controls, aerodynamic performance, pilot protection, and just about every other feature on the X-15, was experimental. For example, the aircraft included the first inertial navigation system (INS), and used a sophisticated stability

---

[2] Robert White, Joseph Walker, Robert Rushworth, John "Jack" McKay, Joseph Engle, William "Pete" Knight, William Dana, and Michael Adams.

1.10   X-15 slung under its mother ship – a similar flight architecture is employed by Virgin Galactic. Courtesy: NASA

augmentation system that helped maintain control of the aircraft. Above the atmosphere, reaction controls kept the aircraft stable, while hand controllers in the cockpit were linked to small hydrogen peroxide thrusters at the nose for pitch and yaw control, and on each wing to control roll. During the descent, the pilot switched from thruster control to traditional stick-and-rudder flying, effectively making the X-15 an unpowered glider (Figure 1.11), much like its descendant, SS1.

One feature SS1 didn't have in common with the X-15, despite it being an experimental vehicle, was an ejection seat; the X-15 pilots could avail themselves of a seat that enabled ejection at speeds up to Mach 4, in any attitude and at any altitude up to 36,500 meters. After the canopy was jettisoned, the seat fired upward and, once clear of the aircraft, deployed a pair of fold-out fins and telescoping booms for stabilization. Oxygen was supplied from two under-seat tanks, and restraints kept the pilot's arms and legs from flailing in the airstream. After reaching a safe altitude, the pilot would unbuckle, jump from the seat, and activate his parachute. Powering the X-15[3] was the LOX and anhydrous

---

[3]   SS1 was fuelled by nitrous oxide, more commonly known as "laughing gas", which is what dentists used to use as anesthesia, and rubber.

1.11   X-15 landing, just like SpaceShipOne. Courtesy: NASA

ammonia-fueled XLR99 rocket engine that generated an awesome half a million horsepower. Like so much of the X-15, the engine was unique because the pilot could use a throttle and it could be restarted.

A typical X-15 flight started with the ground crew disconnecting the servicing carts used to prepare the B-52 and the X-15 for flight. The B-52 then taxied along the dry lake bed before take-off; on a hot day, most of the 3,658 meters of runway was required to achieve lift-off. Twelve minutes prior to launch, the pilot started the auxiliary power units, checked his systems and flight controls, tested the ballistic control system, set the switch positions, activated the propulsion system, and turned on data recorders and cameras. With his preflight checks completed, the pilot prepared for release from the B-52 – a gut-wrenching change from normal 1G to zero-G flight. A few seconds of freefall followed, and then the powerful rocket engine ignited and catapulted the X-15 to Mach 5 and beyond. Climbing at an incredible 1,219 meters per second, the engine shut down after one or two minutes, leaving the X-15 as a super-glider. From sheer momentum, the aircraft continued to climb before orienting itself for an unpowered landing. At around 10,500 meters, the pilot set up his final approach for landing, dumping any remaining propellant to reduce weight, before jettisoning the ventral rudder from the aft underbelly and lowering the landing flaps, nose gear, and rear landing skids. Touchdown would normally be made at a sink rate of 0.6 meters per second and a forward velocity of 320 kilometers per hour, with a slide-out distance of up to 1,800 meters.

The X-15 was a pure research vehicle, whose sole function was to explore the effects of hypersonic travel on man and machine. In the end, the X-15 program yielded a treasure trove of hypersonic lessons learned that proved invaluable to developers of the Space Shuttle and, more recently, to the commercial space vehicle developers such as Scaled Composites.

1.12   WhiteKnight1 and SpaceShipOne. Courtesy: The Spaceship Company

THE LEGACY OF SPACESHIPONE

Fast forward to October 4th, 2004,[4] 36 years after the end of the X-15 program. A momentous event is taking place at Mojave Airport, a sprawling test center in the California desert, a one-hour drive from Los Angeles. Here, at this desolate airport, a small, winged spacecraft built with lightweight composites and powered by a rocket motor using laughing gas and rubber will fly to the edge of space and into the history books. Registered with the Federal Aviation Administration (FAA) by the alphanumeric designation N328KF,[5] but known to space enthusiasts as SS1, this privately developed vehicle (Figure 1.12) will have a galvanizing effect on the commercial spaceflight industry.

The excitement began building the night before, as cars poured into the parking lot and continued to stream in almost until take-off, by which time crowd-control personnel had almost given up. Rows of trucks with satellite dishes and glaring spotlights greet the spectators as they stream into the airport. It's only five in the morning but a sense of expectancy already wafts through the air together with the smell of coffee and bagels. A huge Ansari

---

[4]   The SpaceShipOne team deliberately chose October 4th as the date of their second attempt because of its significance in space history: 47 years earlier, the Soviets had put the world's first satellite, Sputnik 1, into orbit – kicking off the first space race.

[5]   The 'N' in the designation is the prefix used by the FAA for US-registered aircraft and 328KF stands for 328 (kilo – 'K') feet ('F' in the designation), which is the official demarcation altitude for space.

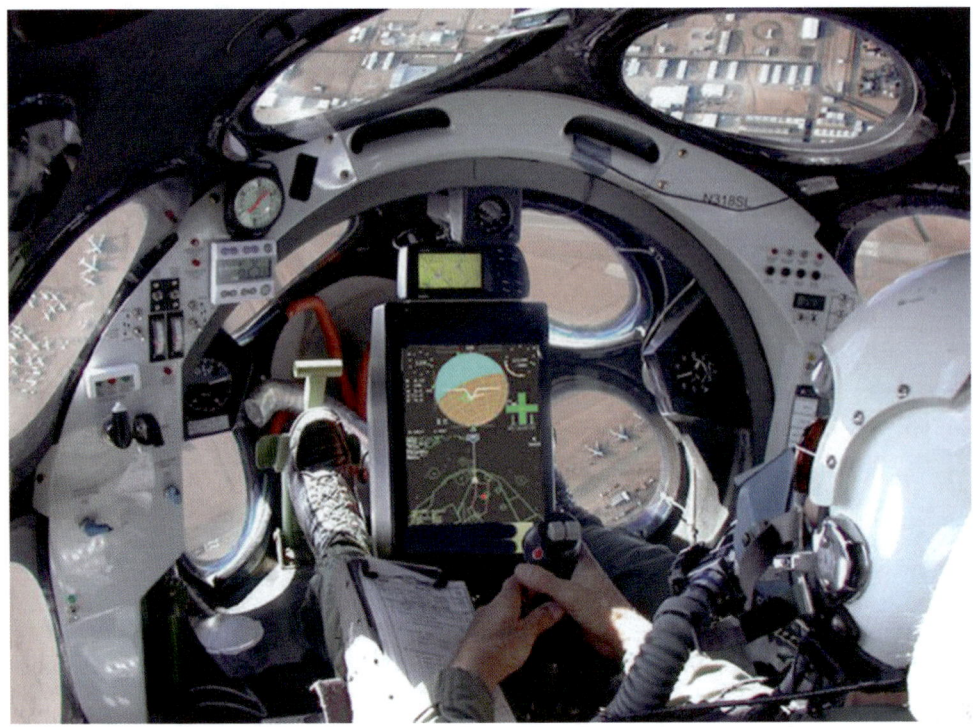

1.13 SpaceShipOne interior. Courtesy: Scaled Composites

X-Prize banner flutters from the control tower as thousands of space enthusiasts from around the world wait for the appearance of N328KF. Buzz Aldrin and other legends of the space program rub shoulders with William Shatner and Mojave's maverick engineering genius, Burt Rutan. Only a few kilometers away, at Edwards Air Force Base on August 22nd, 1963, test pilot Joe Walker reached the edge of space by flying an X-15 to an altitude of 107,333 meters. The X-15 eventually gave birth to the Space Shuttle, a semi-reusable vehicle embroiled in politics that became a symbol that the high frontier was the sole dominion of governments and space agencies – a status quo perpetuated for more than three decades. Until now.

Like all of Rutan's creations, the world's first private spacecraft is an impressive feat of engineering. Marked by simplicity of design, the vehicle doesn't look like it should fly into space. The interior (Figure 1.13) is sparse and devoid of the myriad switches, dials, and toggles that crowd the Shuttle's flight deck. There are a few low-tech levers, pedals, and buttons that suggest the vehicle is designed to fly, but the Spartan design doesn't scream "space".

"White Knight is taxiing" crackles over the public address system. The announcement is followed by the sound of high-pitched jet engines that mark the arrival of a gleaming white carrier aircraft. Slung tightly underneath is SS1. WhiteKnight and SS1 take off followed by two prop-driven chase planes that will follow SS1 during its one-hour ride to separation altitude. "Three minutes to separation" comes the announcement.

1.14 Brian Binnie flies the flag after SpaceShipOne wins the X-Prize. Courtesy: Scaled Composites

Spectators scan the sky searching for the thin white line that is SS1. At the drop altitude of 14 kilometers, SS1 is dropped like a bomb above the Mojave. Falling wings level, ex-Navy Test Pilot and soon-to-be astronaut, Brian Binnie, 51, trims SS1's control surfaces for a positive nose-up pitch. Then he fires the rocket motor, boosting the spacecraft almost vertically. "It looks great," says Binnie as he rockets upwards at Mach 3. Within seconds, SS1 is gone, trailing white smoke. SS1 accelerates for more than a minute, subjecting Binnie to 3 G. At an altitude of 45,000 meters, the engines shut down and SS1 continues on its ballistic trajectory to an altitude of 114,421 meters. A loud cheer erupts from the spectators, who are following the proceedings on a giant screen, each of them euphoric with the realization that high above them is a spacecraft that may one day carry them into space. With his spacecraft's rear wings feathered to increase drag upon re-entry, Binnie prepares to bring SS1 back to Earth. On the ground, the spectators wait, straining to hear the double sonic boom that will announce SS1's return. Seconds later, the unmistakable sound announces that SS1 is on her way back from her historic mission. Binnie (Figure 1.14) guides SS1 gently back to Earth, gliding the spacecraft back

Table 1.2. Manned Suborbital Flights.

Date	Mission	Crew	Country	Remarks
1961-05-05	Mercury-Redstone 3	Alan Shepard	US	First manned suborbital spaceflight; first American in space
1961-07-21	Mercury-Redstone 4	Virgil Grissom	US	
1963-07-19	X-15 Flight 90	Joseph A. Walker	US	First winged craft in space
1963-08-22	X-15 Flight 91	Joseph A. Walker	US	First person and spacecraft to make two flights into space
1975-04-05	Soyuz 18a	Vasili Lazarev Oleg Makarov	Soviet Union	Failed orbital launch; aborted after malfunction during stage separation
2004-06-21	SS1 flight 15P	Mike Melvill	US	First commercial spaceflight
2004-09-29	SS1 flight 16P	Mike Melvill	US	First of two flights to win Ansari X-Prize
2004-10-04	SS1 flight 17P	Brian Binnie	US	Second X-Prize flight, clinching award

to a perfect touchdown on the runway. Welcoming him enthusiastically upon landing are nearly 30,000 spectators, Microsoft's cofounder, Paul Allen, who helped finance the project, Burt Rutan, SS1's designer, and Peter Diamandis, chairman of the Ansari X-Prize Foundation.

In addition to winning the X-Prize, Binnie's flight smashed the altitude record for an airplane, set by X-15 pilot Joseph Walker in 1963 (Table 1.2). Among the VIPs who watched SS1 were Sir Richard Branson, head of the Virgin Group, and Marion Blakey, head of the FAA. After Binnie landed, Blakey presented him with an astronaut pin and paid tribute to him as well as Melvill, the only astronauts to earn their wings from the FAA rather than NASA or the military. The US$10 million was paid, not by the X-Prize Foundation, but by the insurance company the group dealt with – XL Aerospace – in what's known as a "hole-in-one" insurance policy, similar to those taken out by golf courses for tournaments:

> "The Ansari X-Prize is the beginning, it's not the end. Over the course of the last two weeks we have had companies approaching us, we have had wealthy individuals approaching us, about investing in this marketplace. The same thing happened when Lindbergh flew, the same thing happened when Netscape went public, the same thing's going to happen here. Why not have private space travel? Why not be able to climb into a ship and rocket into the sky, and come back and do it again in the afternoon? … Make it accessible to everybody."
>
> Peter Diamandis

REGULATING THE INDUSTRY

To the delight of commercial spaceflight fans, SS1 opened the door for a new industry, but it also highlighted issues about how best to ensure the safety of passengers and crew. Congress acted quickly, passing legislation at the end of 2004, giving birth to the Commercial Space Launch Amendments Act (CSLAA), which gave what most in the industry had sought: removal of the regulatory uncertainty surrounding suborbital spaceflight by assigning jurisdiction over suborbital vehicles to the FAA's Office of Commercial Space Transportation (known by the unusual acronym AST). The legislation also limited the authority AST would have over the industry, allowing companies to mature before the government could enact detailed regulations.

One of the key aspects of the CSLAA was that it limited the ability of AST to issue strict regulations regarding crew and passenger safety until 2012. At the time of the legislation, the government didn't really know how they should regulate suborbital passenger vehicles because of a dearth of flight experience (the same was true at the beginning of 2014 when this book was published). So, while the legislation didn't prevent AST from regulating vehicles to protect the safety of third parties, or the "uninvolved public", it was fairly flexible when it came to issues such as flight training; rather than issuing strict regulations for flight crew operations, the AST released draft guidelines, although government officials made it clear the guidelines were drafts, and that they were open to modifications. Most in the industry welcomed the common-sense AST approach, which was designed to avoid restraining the industry (the word "should", not "must", is used throughout). For example, the flight crew guidelines recommend pilots have a pilot's license, a valid medical certificate, and be thoroughly trained in all aspects of the vehicle's flight systems. Seems reasonable. The passenger guidelines state that companies operating suborbital vehicles should inform passengers of the risks of flight (Appendix II), perform medical examinations of prospective passengers, and provide preflight safety training. Again, nothing wrong with that. The section dealing with medical qualification was a little more specific, identifying how many Gs passengers should be subjected to, and what types of disorders should be screened for, but the guidelines still echoed the common-sense approach of the rest of the document. While space advocates seemed happy with the language set out in the CSLAA legislation, there has been much discussion among lawmakers on the subject of whether the industry can self-regulate, especially since an FAA reauthorization Bill extended the original legislation until October 2015, by which time a few vehicles should have flown.

One of the arguments made by industry proponents against government regulation is that it's not in the interest of companies to kill their passengers. It's a common-sense attitude that is understandable and, for the most part, acceptable, given the outstanding safety records of Virgin and Scaled Composites. But they aren't the only players in the suborbital game, and new companies may not be as committed to passenger safety, hence the argument for greater regulations. To that end, one of the initiatives in recent years has been the creation of a Voluntary Personal Spaceflight Industry Consensus Standards Organization to develop standards for the safe operation of passenger-carrying suborbital spacecraft, thereby obviating the need for government regulations.

Of course, even if more rigorous standards were imposed, there is no such thing as absolute safety, which is why the CSLAA requires that, before receiving compensation from a spaceflight participant or making an agreement to fly a spaceflight participant, an sRLV operator shall inform the spaceflight participant in writing that the US Government has not certified the launch vehicle as safe for carrying crew or spaceflight participants. Why the written consent? Well, first of all, without such reference, in any litigation following an accident, the operator would have difficulty defending its vehicle risk level and demonstrating the thoroughness of the information passed to the customer and the fleet would likely be grounded. Secondly, the government reasons that the average spaceflight participant can't be expected to have the necessary background and technical experience to truly grasp the dangers of spaceflight, which leads us on to a section that describes the risks involved in this suborbital spaceflight business.

SUBORBITAL RISKS

Decompression

We'll start by assessing a couple of the risks common to the airline passenger and spaceflight participant: cabin pressure and decompression. We all know air pressure decreases with increasing altitude, and is close to zero in space. You may also know that when it comes to designing spacecraft (and aircraft) cabins the lower the cabin pressure the better, because this reduces cabin weight. Of course, passengers need some pressure so they can breathe, so a compromise is necessary. Until 1957, commercial airline cabins were required to be pressurized to a maximum equivalent altitude of 3,048 meters (523 mmHg), after which they were required to maintain a maximum allowed altitude of 2,438 meters (564 mmHg). The latest aircraft have cabins pressurized to about 1,524 meters because research showed that the 2,438-meter limit might have an adverse effect on some passengers. For example, one study found that a substantial proportion of healthy passengers aged 65 years or more would have inadequate arterial oxygen levels while breathing air at that altitude. This means the designer of a suborbital spacecraft will have to design a cabin that can be pressurized to the highest equivalent altitude possible considering the demographic characteristics of the paying passengers.

Another problem facing the spacecraft designer is protecting the passenger in the event of a rapid decompression (Figure 1.15). If such an event were to occur in space, cabin pressure would be lost quickly. Very quickly. Decompression to about 7,620 meters leaves a healthy normal young person conscious for about six minutes. This time of useful consciousness (TUC) decreases to just 15 seconds at 13,716 meters. Death follows shortly thereafter. To deal with a rapid decompression event, passengers can either breathe supplemental oxygen or the cabin can be pressurized. Or both. The problem with either of these solutions is *weight* because an oxygen subsystem must be integrated into the spacecraft cabin. Also, breathing pure oxygen is limited to certain altitudes. At 10,058 meters, breathing pure oxygen is about the same as breathing air at sea level. Above 12,192 meters, 100% oxygen must be under positive pressure to maintain the equivalent altitude of 3,048 meters and, at altitudes above 15,240 meters, a passenger requires a pressurized suit to be safe. At 16,764 meters, atmospheric pressure is so low

1.15 The aftermath of a rapid decompression. On July 25th, 2008, a Qantas 747–400 (VH-OJK, Flight 30) experienced a rapid decompression while cruising at 8,840 meters after an oxygen cylinder that was part of the emergency oxygen system exploded and blew a hole in the fuselage. Imagine this happening in a spacecraft. Courtesy: Qantas

that water vapor in the body appears to boil, causing the skin to inflate like a balloon, and, at 19,202 meters, blood at normal body temperature (36.6°C) appears to boil. One solution to avoid most of these problems is to wear a pressure suit. These used to be expensive items, costing more than US$1 million, but companies like Final Frontier Design (Figure 1.16) will be offering their suits for around US$20,000. More about them later in the book.

The average effective performance time (EPT) for aircrew without supplemental oxygen is as follows:

4,572–5,486 meters:	30 minutes or more;
6,706 meters:	5–10 minutes;
7,620 meters:	3–5 minutes;
8,534 meters:	2–2.5 minutes;
9,144 meters:	1–2 minutes;
10,668 meters:	30–60 seconds;
12,192 meters:	15–20 seconds;
13,716 meters:	9–15 seconds.

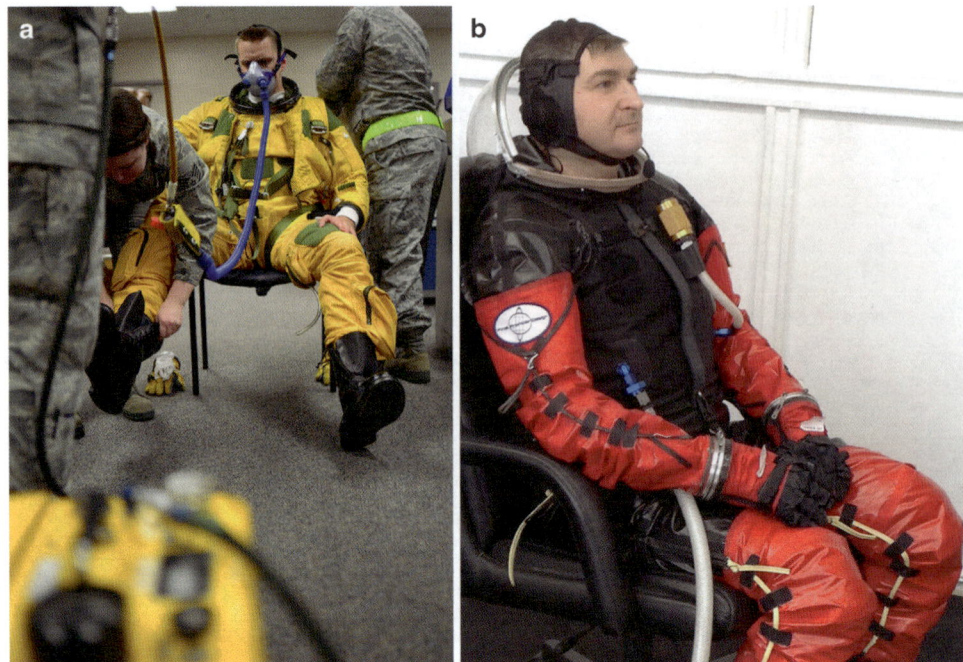

1.16 (**a**) At least one commercial spaceflight company is considering the use of pressure suits for its passengers. A survey of U-2 pilots found that more than three-quarters reported symptoms of decompression sickness during their careers – more than 10% of the pilots reported they altered or aborted their missions as a result. The U-2 cabin was designed to maintain an equivalent altitude of 8,840–9,144 meters, with the pilot wearing a pressure suit capable of maintaining an altitude of 10,668 meters and breathing pure oxygen during the flight and for a one-hour period prior. Courtesy: NASA. (**b**) Nic of Final Frontier Design (FFD) wears an affordable pressure suit. FFD will sell them for around US$20,000, compared to the US$100,000-plus cost of other pressure suits. Courtesy: FFD

Factors that determine EPT are:

1. Altitude: EPT decreases at high altitudes.
2. Rate of ascent: the faster the rate, the shorter the EPT.
3. Physical activity: exercise decreases EPT considerably.
4. Day-to-day factors: physical fitness and other factors (smoking, health, stress) may affect ability to tolerate hypoxia.

One of the drawbacks of wearing a pressure suit is that it requires extra training and extra time because of the need to complete an oxygen pre-breathe to reduce the risk of decompression sickness (DCS). DCS is a condition caused by dissolved gases coming out of solution and forming bubbles in the body during depressurization. Since bubbles can form in any part of the body, DCS can produce effects such as joint pain, paralysis, and death. But, while a pre-breathe reduces the incidence of DCS, it is logistically complicated and expensive, and there is still the risk of DCS (Figure 1.17).

DCS Type	Bubble Location	Signs & Symptoms (Clinical Manifestations)
BENDS	Mostly large joints of the body (elbows, shoulders, hip, wrists, knees, ankles)	• Localized deep pain, ranging from mild (a "niggle") to excruciating–sometimes a dull ache, but rarely a sharp pain • Active and passive motion of the joint aggravating the pain • Pain occurring at altitude, during the descent, or many hours later
NEUROLOGIC Manifestations	Brain	• Confusion or memory loss • Headache • Spots in visual field (scotoma), tunnel vision, double vision (diplopia), or blurry vision • Unexplained extreme fatigue or behavior changes • Seizures, dizziness, vertigo, nausea, vomiting, and unconsciousness
	Spinal Cord	• Abnormal sensations such as burning, stinging, and tingling around the lower chest and back • Symptoms spreading from the feet up and possibly accompanied by ascending weakness or paralysis • Girdling abdominal or chest pain
	Peripheral Nerves	• Urinary and rectal incontinence • Abnormal sensations, such as numbness, burning, stinging and tingling (paresthesia) • Muscle weakness or twitching
CHOKES	Lungs	• Burning deep chest pain (under the sternum) • Pain aggravated by breathing • Shortness of breath (dyspnea) • Dry constant cough
SKIN BENDS	Skin	• Itching usually around the ears, face, neck, arms, and upper torso • Sensation of tiny insects crawling over the skin • Mottled or marbled skin usually around the shoulders, upper chest, and abdomen, accompanied by itching • Swelling of the skin, accompanied by tiny scar-like skin depressions (pitting edema)

1.17 Signs and symptoms of altitude decompression sickness. Courtesy: Flyngo

If you fly on commercial aircraft you no doubt are familiar with the safety briefing that explains how the oxygen mask will drop in front of you in the event of loss of pressure. The supply of oxygen is good for about 8–10 minutes because commercial aircraft are only a short time away from survivable atmospheric air pressure in such an event. It's a different story if you happen to be flying on a suborbital vehicle because a suborbital spacecraft is committed to the ballistic part of its trajectory from the final phase of the rocket motor burn until it gets back down to breathable (survivable) air. This could take several minutes in the event of a depressurization event during the burn and, depending on when the depressurization was to occur, this period of time would be way beyond the ability of a passenger to survive without a pressure suit.

Acceleration

Another consideration is acceleration. Acceleration (G) is one of the most demanding aspects of flying in a suborbital spacecraft because of the risk of gravity-induced loss of consciousness (G-LOC – pronounced "gee-lock"). To better understand why G-LOC poses such a threat to passenger safety, it's helpful to understand what happens to the body when subjected to positive G (abbreviated +Gz). The most significant effect of +Gz on the brain and the eyes is a reduction in blood pressure and blood flow. The eyes react first. As G-forces increase and blood pressure drops, passengers will experience grayout (loss of color and clarity), tunneling of vision, and blackout. If Gs continue to increase beyond

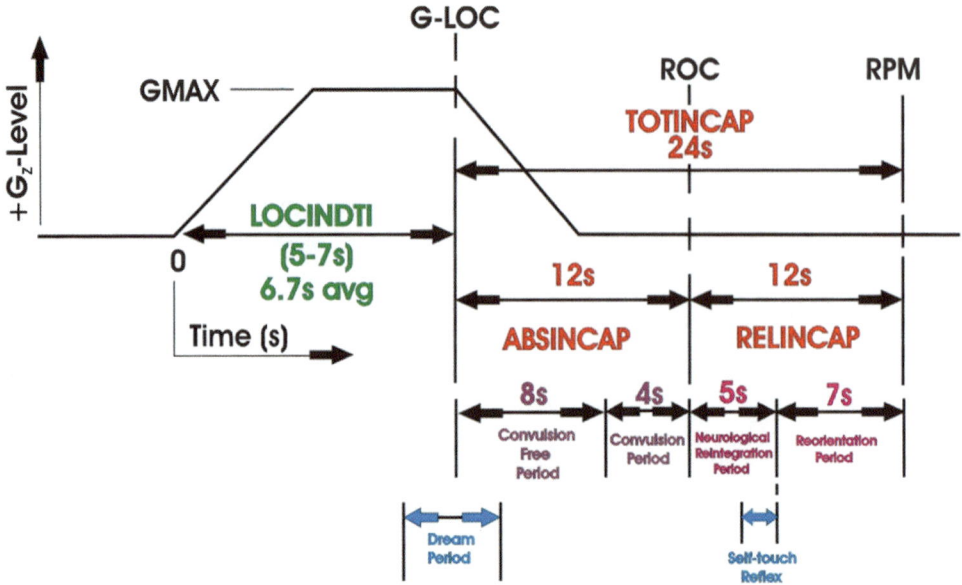

1.18 G-LOC time relationship. LOCINDTI = loss of consciousness induction time. ROC = Return of consciousness. RPM = Return of purposeful movement. ABSINCAP = absolute incapacitation period. RELINCAP = relative incapacitation period. TOTINCAP = total incapacitation. Courtesy James E. Whinnery Ph.D., M.D., Aeromedical Research Division Civil Aerospace Medical Institute/NASA

blackout, the passenger will G-LOC. That's because when the brain loses its blood supply and exceeds its oxygen reserve, it abruptly fails. Once it fails, it stays "turned off" for a variable length of time, even after blood flow is restored. Consciousness can be maintained when the G-onset rate is slow enough (say 1 G per second) that visual symptoms are recognized but, when G-onset rates are high (5 or 6 G per second), and the peak G-level sustained is high, G-LOC can occur without any visual warning signs. G-LOC is a serious problem because no one is immune.

Once a passenger G-LOCs, their brain enters a state called *absolute incapacitation*, which lasts for about 15 seconds (the range is anything from 5 to 30 seconds). This state will occur even if the G is unloaded. While incapacitated, the passenger will be in a dream-like state, unaware of their environment and unable to respond to any outside stimuli. The brain's blood supply returns during this period and gradually the brain starts to "wake up" and enters a period of *relative incapacitation* which lasts another 12–15 seconds. The combined absolute and relative incapacitation times are referred to as the *total incapacitation time*. At the end of this total incapacitation period, the passenger will be able to recognize where they are and respond to the environment. The third phase of recovery from G-LOC is the return of cognitive processing skills which may require several minutes before a return to full function (Figure 1.18) is realized. During this period, functional motor skills and situational awareness may be severely impaired.

How will the passenger feel when they recover from G-LOC? First of all, they may not even recognize the G-LOC incident because one of the symptoms of G-LOC is partial amnesia caused by impaired oxygen flow to the brain. As they gradually recover from the discombobulating feeling of regaining full consciousness, they may also experience tingling around the mouth or in the extremities and perhaps a sense of dreaming – part of a constellation of symptoms which are defined collectively as the G-LOC syndrome:

- Loss of peripheral vision
- Tunnel vision
- Blackout (complete loss of vision)
- Loss of consciousness
- Loss of motor control (purposeful movement)
- Loss of sensory input to the brain
- Lack of memory formation
- Myoclonic convulsions
- Dreamlets
- Recovery of consciousness
- Neurological reintegration
- Neurological external environment reorientation
- Return of purposeful movement
- Transient tingling or slight numbness of the extremities
- Alteration of psychological state (anxiety, confusion, embarrassment).

While a G-LOC event may cause our passenger to feel a little queasy, the syndrome is a protective mechanism that has evolved to protect us in a gravitational field and to ensure the optimum protection of the organ system that is the key to its evolutionary success on Earth: the brain. And it's not as if the brain gives up at the slightest hint of G-stress. Far from it. As soon as G-stress is detected, the cardiovascular and neurological systems initiate protective reflex mechanisms, so functional compromise does not occur easily. With the passenger unconscious, the brain is in a minimal energy expenditure state, with loss of sensory, motor, and consciousness function. If a G-LOCed passenger were to be hooked up to an electroencephalogram (EEG), a flight surgeon would observe a synchronized slow wave pattern that would persist until the recovery process started – the relative incapacitation period. At this stage, blood flow would begin to return to normal levels and *myclonic* twitching might occur. You may wonder why the passenger would start convulsing, but the mechanism, like all the G-LOC recovery processes, has a purpose. The twitching serves to contract the muscles in the extremities and abdomen, thereby enhancing return of blood to the central circulation and ultimately the brain.

+Gz also exerts mechanical effects on soft tissues and compresses the spine and it affects the cardiovascular and pulmonary systems, creating visual symptoms and the risk of G-LOC. −Gz is just as bad, as it causes visual and cardiovascular disturbances and is just as capable of causing G-LOC. Then there are the mechanical effects of acceleration, which causes soft tissues to sag, with the result that a person subject to G appears to have aged prematurely. Fortunately, it's a reversible change. The bottom line is that the sheer magnitude of G in any axis causes problems. Above 2.5 G, most people find it difficult to

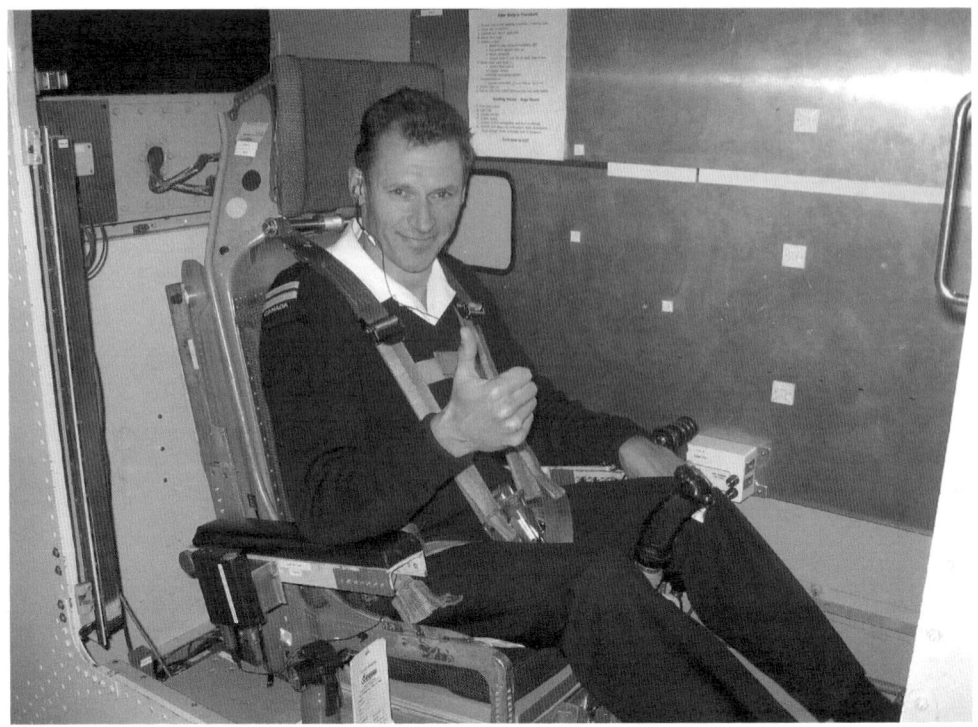

1.19 The author seated in Canada's only manned centrifuge (since decommissioned), waiting for his run. As Director of Canada's Manned Centrifuge Operations I was able to take a ride whenever I wanted. Courtesy: Chris Townson

rise from a seat and, when that acceleration increases to 3 G or more, raising an arm is a workout. Crank up the Gs to +8G and any gross movement is next to impossible. Even if your name is Arnold Schwarzenegger.

In tandem with the mechanical effects are the hydrostatic effects, since acceleration increases the weight of the blood, thereby increasing the pressure gradient in the hydrostatic column, which creates havoc in the cardiovascular system. +Gz also makes breathing difficult by pulling down the diaphragm and collapsing the air sacs in the lungs (the greater the G, the more air sacs collapse, like a balloon collapsing) causing a G-induced symptom known as *acceleration atelectasis*.

There are myriad factors that combine to determine G-tolerance on a given day. For example, short passengers have a higher tolerance than tall passengers due to the respective differences in heart-to-brain distances. Individuals with higher blood pressure have a higher tolerance than those with lower blood pressure. Being sick with dehydration can also reduce G-tolerance. Acclimatization training is also an important factor in G-tolerance. For example, suborbital passengers are required to undergo centrifuge (Figure 1.19) high-G training to ensure they are proficient in performing protective anti-G straining maneuvers (AGSM), which we'll discuss later. Finally, a good understanding of factors relating to G-tolerance, especially the G-LOC syndrome, is vital for all who enter the high-G environment. Having said that, there are certain factors that may limit an individual's ability to tolerate G.

An abnormality of the neurological or neurovascular system will most likely preclude an individual from engaging in high-G activity due to the potential for sudden incapacitation, since any such abnormality might contribute to compromising blood supply to the brain. Since the cardiovascular system is the system primarily affected by +G$_z$, any abnormality in cardiovascular anatomy or physiology is reason for concern in aerospace safety. Equally, medications that alter cardiovascular physiology are viewed with caution, especially pharmacological agents that alter blood pressure and the function of the heart. The heart is particularly susceptible because acceleration is a *dysrhythmogenic* stress which means that anything that affects cardiac rate or rhythm is a threat to safety. Flight surgeons have a name for the group of symptoms that may affect the heart during +Gz: *tachydysrhythmias*. Like most medical terms, *tachydysrhythmia* appears to have been borrowed from an alien language but, in layman terms, it simply means a quickening of heart rate (ventricular tachycardia) and premature beats (supraventricular and atrial) which are common during +G$_z$. For the symptoms that occur following +G$_z$, flight surgeons have another tongue-twister: *bradydysrhythmias*. Like its counterpart, this term has a simple explanation, describing the out-of-sequence beats (sinus arrhythmia), bradycardia, and spontaneous heart beats (ectopic atrial rhythm) that occur following +G$_z$.

Musculoskeletal problems, particularly those affecting the neck and back, are of particular concern during +G$_z$ stress. For example, any anatomical abnormality that decreases neck or spinal strength or stability has to be carefully considered. It's the reason pilots are prescribed neck and back muscle-strengthening exercises to prepare themselves for the G environment. The pulmonary system is also significantly affected by +G$_z$ stress; blood can be drawn away from the lungs, resulting in less oxygen being delivered to the muscles and to the nervous system.

Tolerance limits for +Gz acceleration are usually signaled by visual symptoms such as peripheral light loss (PLL), tunneling, grayout, and blackout. During gradual onset of acceleration, a relaxed subject not wearing an anti-G-suit typically experiences initial visual symptoms at about +4 Gz, although susceptible individuals may experience PLL as low as +2 Gz. "G monsters", on the other hand, may not notice anything until +7 Gz. In my job as director of Canada's manned centrifuge operations, I witnessed about half a dozen such individuals, one of whom casually chatted away as the G-level crept over the 7-G mark! For most people, once they've experienced initial visual symptoms, the next symptom will likely be blackout at about +5 Gz and unconsciousness if the Gz continues to increase. Once the individual has lost consciousness, they are deemed to have G-LOCed. Witnessing an episode of G-LOC can be a little disconcerting. First, the hapless individual's head drops to the chest and seizure-like flailing motions may occur – some people have broken limbs as a result of this. While their limbs are flailing, the now-unconscious individual may exhibit myoclonic, spastic-like twitching. Fortunately, consciousness returns quickly (a typical G-LOC period is 15 seconds) and the individual will slowly raise their head, looking very, *very* confused. This confusion is quickly exacerbated when they realize they don't remember the incident and, when asked whether they know what happened, some even deny losing consciousness (I once argued with a pilot who insisted he hadn't G-LOCed until I showed him video evidence to the contrary).

So, how do you reduce the effects of G? There are two approaches; the first is to decrease the vertical distance between the heart and brain by tilting the seat back, and the second is to apply counter pressure against the legs and abdomen to reduce blood pooling there. The counter pressure can be generated by performing the AGSM and/or by wearing a G-suit. Considering the G envelopes of the current crop of suborbital vehicles, these measures should be enough for most people. Most. One of the biggest medical disqualifiers in suborbital flight will probably be related to cardiovascular problems exacerbated under G, and here's why. Heart rate increases and the vascular return pressure is *reduced* (decreasing preload) under acceleration, which means the heart is starved by decreasing filling of the atria during diastole. This isn't a problem in passengers with a healthy cardiovascular system because the muscle straining associated with the AGSM and activation of an anti-G-suit will increase resistance to circulation, which increases the systolic pressure. But, if a passenger has some kind of cardiovascular abnormality (an arrhythmia, for example), this sequence of events may not happen in exact synchrony with G-loading, with the result that the heart after-load (peripheral flow resistance) will fluctuate – possibly out of control. Now you may be thinking that passengers with suspect hearts won't be allowed to fly because these arrhythmias will be detected in the medical exam. Not so. That's because the effects on the heart that were just described are not always detected on an electrocardiogram (ECG), whether the ECG is conducted under resting or exercise conditions. A treadmill exercise protocol is a great way of assessing whether someone can run a marathon but, because such a test doesn't drop preload or cardiac output, and because peripheral resistance to blood flow is decreased and output from the heart increases with exercise on a treadmill, this test can't assure that a passenger can survive prolonged G-stress.

On the subject of heartbeat abnormalities (arrhythmias) that occur during acceleration, it's worth pointing out that in a series of 1,180 centrifuge training sessions involving professional aeromedical attendees at the USAF School of Aerospace Medicine, 47% resulted in arrhythmias. Of these, 4.5% should have resulted in termination of the session. The point is that these *arrhythmias can occur in pre-screened individuals*; the arrhythmias might be harmless for the person in the street but, when subjected to G-load, the abnormality could be lethal. At least in a centrifuge, if a passenger were to develop an electrocardiographic abnormality during a centrifuge run, the centrifuge can be stopped, but if the problem manifests itself during a suborbital flight the result may be a dead passenger; and, don't forget, if this happens during the boost phase, the vehicle is committed to the ballistic phase of the flight.

Bail-Out

Now let's consider a vehicle with an emergency egress capability. Imagine if something goes wrong shortly before re-entry and the lone passenger (this very rich passenger wanted the flight to himself) is instructed to bail out. From 100,000 meters! Would this person survive? Perhaps. Perhaps not. What *is* certain is that the individual would be in for a memorable ride. Here's what might happen. For the first 70 seconds, the astronaut would descend to about 80,000 meters and accelerate to about 700 meters per second. As the descent continued, acceleration would continue up to about 1,000 meters per second (Mach 3.1) at 114 seconds into the fall. At this point, the astronaut is at an altitude of about 45,000 meters and finally begins to decelerate, until crossing 9,000 meters 243 seconds into the fall at a speed of about 88 meters per second; 325 seconds into the bail-out, the astronaut would cross 3,000 meters at 63 meters per second. During the bail-out, 22

1.20 Felix Baumgartner waits for the moment of truth. Supersonic skydiver Felix Baumgartner was faster than he or anyone else thought during his record-setting jump in October 2012 from 38,624 meters up. "Fearless Felix" reached 1357.64 kilometers per hour, equivalent to Mach 1.25, becoming the first human to break the sound barrier with his body. He wore a pressurized suit and jumped from a capsule lifted by a giant helium balloon over New Mexico. His free fall lasted four minutes, 20 seconds. Fifty-two million people watched YouTube's live stream of the exploit. Courtesy: Red Bull Stratos

seconds would be spent above 2 G, 18 seconds above 3 G, 13 seconds above 4 G, and 6 seconds above 5 G. Maximum acceleration would be about 5.8 G at 137 seconds into the fall and an altitude of about 26,500 meters. In addition to the acceleration forces, our bail-out victim would also have to contend with extreme cold and the problem of maintaining stability. This latter factor is probably the most dangerous of all, especially during the high-acceleration phases of entry into denser atmosphere because if the astronaut enters a flat spin blood is centrifuged into the extremities and blackout is the result. This is what happened to intrepid sky-diver Felix Baumgartner when he made his record-breaking jump from the stratosphere (Figure 1.20). Baumgartner went into a dreaded flat spin while still supersonic and spun for 13 seconds at approximately 60 revolutions per minute, making 14–16 spins before managing regaining control. As sky-diving legend Joe Kittinger noted after the jump, if a highly trained jumper like Baumgartner with 2,500 jumps couldn't prevent a flat spin, "an astronaut, pilot or space tourist could not overcome this spinning probability".

Vehicle Design

Staying on the subject of survival strategies, just how risky are the different designs of sRLVs and how will this influence your decision of which operator to fly with? One consideration you may want to chew over is the potential failure mode of the vehicle. For example, you may want to consider your options if the sRLV is a vertical take-off/vertical landing (VTVL) vehicle. In this case, if a motor fails or is shut down during the first few seconds of flight, the vehicle will be lost. But, if a horizontal take-off sRLV suffers a motor shutdown during the first few seconds of flight it may be able to initiate a runway abort or a go-around procedure. Another matter to deliberate is whether the vehicle has one engine or a cluster of engines. If it uses a cluster, engine-out capability may exist, but the probability of a motor failure for a multiengine cluster is greater than for a single motor of similar reliability. So, if a single motor is 99.9% reliable, the probability of a motor failure in a single mission is 0.1%. But, for a five-motor cluster, the probability of a failure involving at least one motor is about 0.5%. Decisions, decisions.

Emergency Egress Options

Another concern you have probably mulled over is your emergency egress option. What are your chances of getting out of the vehicle if everything goes pear-shaped? Does the vehicle have ejection seats? Probably not. That's because ejection seats are deadweight during a very-low-altitude abort because the vehicle will most likely be lost and the crew and passengers must be transported clear of an almost fully fueled vehicle's fireball. To escape from this scenario, the vehicle needs some kind of rocket-powered escape capsule. Once you gain altitude, there is some limited usefulness for ejection seats in non-catastrophic vehicle failures (explosions), although they become even more limited at about Mach 3.7 at 20,000 meters altitude because of high dynamic pressures. So it's unlikely your vehicle will have ejection seats. So, what do you do in the event of a decompression or a failure of the life-support system? Or both? Well, after the propulsion burn, a sRLV is committed to a ballistic trajectory for several minutes, so staying with the vehicle is probably your best survival strategy. As long as you're wearing a pressure suit. Assuming you survive a non-catastrophic event, chances are the vehicle may not be able to navigate to the planned landing site. Perhaps you'll end up experiencing a post-abort water landing. Does your operator train you for sea survival (Figure 1.21) and emergency egress from an upturned vehicle pitching up and down in a rolling sea? If not, perhaps you should invest in some training. And what if the vehicle is filled with smoke and flames post landing? Does the operator train you for emergency egress under such conditions? If not, why not?

Radiation

Next is radiation. When we talk about harmful radiation in space, we're usually referring to *ionizing radiation*. This type of radiation consists of subatomic particles that can interact with biological tissues and destroy DNA strands, causing genetic damage that can in

1.21   Sea survival training. Courtesy: A4H

turn lead to dangerous mutations. The sources of ionizing radiation in space are galactic cosmic radiation (GCR), solar radiation, solar flares, and the trapped radiation from the Van Allen belts. GCR originates outside of the Solar System and consists of hydrogen nuclei protons (87%), helium nuclei alpha particles (12%), and damaging high-energy heavy nuclei such as iron (1%), while solar cosmic radiation (SCR) comprises proton– electron plasma ejected from the surface of the Sun. Completing the radiation cocktail are solar flares, which are magnetic disturbances on the Sun's surface, generating electromagnetic radiation of the Van Allen belts, which contain trapped protons, heavy ions, and electrons. How much radiation is too much? Well, the dose standard for radiation exposed workers is 20 mSieverts (Sv) a year (averaged over five years); exposure to this level over 40 years results in an excess lifetime fatal cancer risk of 3.2%. By comparison, orbital spaceflight results in a variable radiation dose exposure dependent on orbital altitude and solar activity and ranges from 0.01 to 0.1 Sv per month. NASA astronauts have career exposure limits based upon a maximum of 3% excess lifetime cancer mortality; these limits are recommended by the National Council on Radiation Protection (NCRP). Radiation levels at a suborbital flight altitude would be similar to high-altitude Concorde flights, which mean commercial suborbital astronauts should receive less than 15 microSv per hour during their flight. To put this in perspective, you would need to fly more than 10 suborbital flights to experience the same radiation dose as you get from an X-ray. So don't worry too much about radiation.

Still interested in flying a suborbital mission? It sounds risky, which is why some people find it hard to understand why the industry isn't more heavily regulated. Commercial suborbital spaceflight argues that excessive regulation would kill the industry, but, for experienced space safety professionals, the opposite is true. FAA/AST Associate Administrator, George Nield, a veteran of aviation safety, shares that opinion and, at the 2008 Space Frontier Foundation's annual meeting, he said it was time that space tourism operators got real about the extraordinary risks they faced, warning that, while many companies were presenting their efforts as pioneering a "golden age", neglecting safety could mean "an end to commercial human space flight before it has chance to get started".

2

Suborbital Market

The suborbital spaceflight market represents a fundamental shift in the nature of the manned spaceflight business. When SpaceShipTwo's and Lynx's start routine suborbital flights, it will jumpstart a new commercial space industry. But what exactly is this industry and just how viable and successful will it be?

In November 2011, Space Florida – the State of Florida's spaceport authority and aerospace economic development agency – and the Federal Aviation Administration Office of Commercial Space (FAA-AST) partnered to commission a study prepared by the Tauri Group to forecast the 10-year demand for suborbital reusable launch vehicles (sRLVs). The analysis interviewed 120 potential users and providers, polled 60 researchers, assessed budgets, and surveyed more than 200 high-net-worth individuals. The results of the study – Suborbital Reusable Vehicles: A 10-Year Forecast of Market Demand – were made available to the public via the Space Florida and FAA websites and, since it's the only study of its kind, much of what appears in this chapter is taken from the study.

At the heart of the survey – and the new spaceflight industry – are the sRLVs (Figure 2.1 and Table 2.1), commercially developed vehicles capable of carrying passengers and/or cargo. At the time of writing, 11 sRLVs are in active planning, development, or operation, by six companies. The payload capacity of these vehicles ranges from tens to hundreds of kilograms, and some vehicles such as SpaceShipTwo (SS2) can carry passengers. Others, like the Lynx, will also launch very small satellites. The companies developing these vehicles are ambitious and hope to fly regularly, and by regularly we're not talking about once a month; these vehicles may fly several times a day. Given these lofty goals, it made sense to provide information to the government and industry about the potential of this breed of space vehicle, and that's what the Tauri Group did. By analyzing dynamics, trends, and areas of uncertainty in the eight distinct markets (Table 2.2) sRLVs could support, the group came up with a projected demand.

E. Seedhouse, *Suborbital: Industry at the Edge of Space*, Springer Praxis Books, DOI 10.1007/978-3-319-03485-0_2, © Springer International Publishing Switzerland 2014

2.1   Suborbital reusable launch vehicle. Courtesy: EADS

Table 2.1   SRLVs in Development.

Company	SRLV	Seats	Locker equiva-lents	Cargo (kg)	Price	Operational date
UP Aerospace	SpaceLoft XL		0.5	36	US$350k	2006 (actual)
Armadillo	STIG A		1	10	per launch	2012
Aerospace	STIG B		2	50	Not announced	2013
	Hyperion	2	12	200	US$102k/seat	2014
XCOR	Lynx Mark I	1	3	120	US$95k/seat	2013
Aerospace	Lynx Mark II	1	3	120	US$95k/seat	2013
	Lynx Mark III	1	28	770	US$95k/seat	2017
Virgin Galactic	SpaceShipTwo	6	36	600	US$200k/seat	2014
Masten Space	Xaero		4	25	Not announced	2013
Systems	Xogdor					
Blue Origin	New Shepard	3+	5	120	Not announced	Not announced

To generate as accurate a picture of sRLV market dynamics as possible, the Tauri Group forecast demand for each market (Table 2.3) based on three scenarios:

- *Baseline scenario*: sRLVs operate in a predictable political/economic environment. Existing trends generate demand for sRLVs.
- *Growth scenario*: New dynamics emerge from marketing, branding, and research successes. Commercial human spaceflight has a transformative effect on consumer

Table 2.2 SRLV Markets.

COMMERCIAL SPACEFLIGHT	BASIC AND APPLIED RESEARCH
Human spaceflight experiences for tourism or training	Basic and applied research in various disciplines
AEROSPACE TECHNOLOGY TEST AND DEMONSTRATION	MEDIA AND PUBLIC RELATIONS
Aerospace engineering to advance technology maturity or achieve space demonstration, qualification/certification	Using space to promote products, increase brand awareness, or film space-related content
EDUCATION	SATELLITE DEPLOYMENT
Providing opportunities to K-12 schools, colleges, and universities to increase access to and awareness of space	The use of SRLVs to launch small payloads into orbit
REMOTE SENSING	POINT-TO-POINT TRANSPORTATION
Acquisition of imagery of Earth and Earth systems for commercial, civil government, or military applications	Future transportation of cargo or humans between different locations

Table 2.3 Seat/Cargo Equivalents.

Scenario	*Year*										
	1	*2*	*3*	*4*	*5*	*6*	*7*	*8*	*9*	*10*	*Total*
Baseline	373	390	405	421	438	451	489	501	517	533	4,518
Growth	1,096	1,127	1,169	1,223	1,260	1,299	1,394	1,445	1,529	1,592	13,134
Constrained	213	226	232	229	239	243	241	247	252	255	2,378

behavior, and more customers buy flights. Research results are productive and attract new government and commercial interest.

- *Constrained scenario*: sRLVs operate in an environment of dramatic reduction in spending, due, for example, to worsened global economy.

The Tauri Group also compared forecasts for all markets by scenario, as shown in Figure 2.2, which indicated demand for sRLVs is dominated by commercial human spaceflight, based on an analysis of 8,000 high-net-worth individuals who are sufficiently interested in buying a ticket.

As you can see in Table 2.3, the baseline forecast is 373 seats in Year #1, growing to more than 500 seats by Year #10, totaling more than 4,000 over the first decade. But, if commercial spaceflight really takes off, the growth scenario predicts 13,000 seats. If, on the other hand – and let's hope this doesn't happen – the industry stalls, the constrained scenario predicts just over 2,000. Of course, as meticulous as the Tauri Group was in generating their forecast, there is still uncertainty due to the dynamics of demand as it responds to future events. For example, demand may not always be steady because it could grow more rapidly than predicted based on social dynamics following successful launch

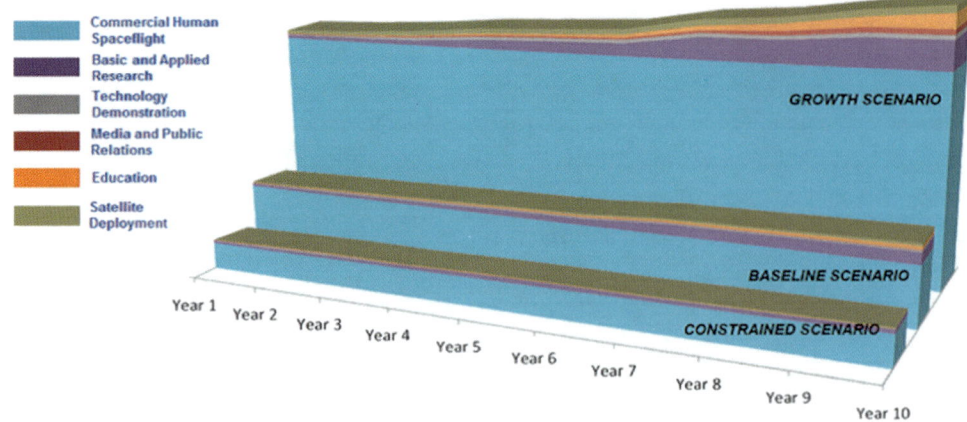

Commercial Human
Spaceflight

Basic and Applied
Research

Technology
Demonstration

Media and Public
Relations

Education

Satellite
Deployment

GROWTH SCENARIO

BASELINE SCENARIO

CONSTRAINED SCENARIO

Year 1
Year 2
Year 3
Year 4
Year 5
Year 6
Year 7
Year 8
Year 9
Year 10

2.2   Potential growth in the suborbital markets. Courtesy: Tauri Group

experiences. Equally, if a vehicle suffers a major malfunction with loss of life, demand would decline sharply and probably come to a standstill. The bottom line is that the forecast is presented as a relatively steady state in each scenario, reflecting current levels of interest.

Going back to Figure 2.2, we can see the second largest area (about 10%) of demand is basic and applied research, funded by government agencies, and research not-for-profits, universities, and commercial firms. Given all the media coverage generated by the subject of space tourists, it's easy to forget sRLVs will be used for science missions, just as they will be utilized to perform technology test and demonstration and satellite deployment. But, the majority of sRLV demand will come from individuals and, because this market is a consumer market, the capability and viability of sRLV ventures will be influenced by individual decision makers. What this means is that, unlike enterprise users (whose demand is for cargo rather than seats) who often have decision-making lead times measured in years, individuals can make purchasing decisions quickly and the behavior of consumers in the industry is unknown. It is this unknown factor that makes it difficult to estimate revenue because consumer behavior can only be based on assumptions – will passengers fly only once or will they be repeat customers? Equally, if sRLV capabilities vary from current expectations, levels of activity could be higher or lower. For example, NASA, the Canadian Space Agency (CSA), and the European Space Agency (ESA) might decide to use sRLVs for astronaut training and research integrated with International Space Station (ISS) activities. This development isn't too far-fetched given that the forecast predicts more than 50 governments will begin to fund sRLV research.

In short, the Tauri Group concluded that, at a minimum, demand for suborbital flights will be sustained, and be sufficiently robust to support multiple providers with a baseline demand over 10 years exceeding US$300 million in revenue. But, in the event of increased marketing, research successes, greater consumer uptake, *and* multiple flights per day, revenue could generate US$1.6 billion in the first decade. That's a healthy industry (Figure 2.3).

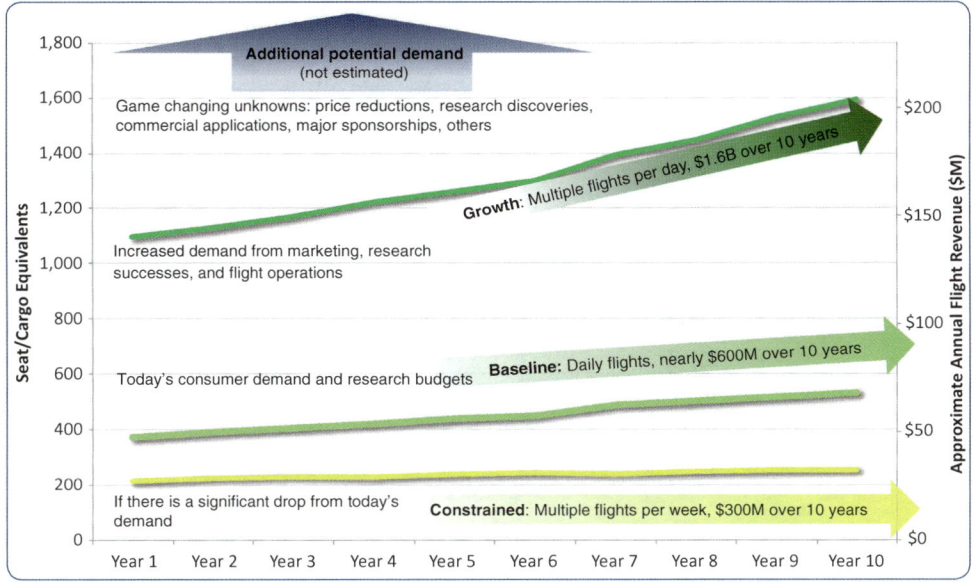

2.3 Potential flight revenue in the suborbital markets. Courtesy: Tauri Group

THE 10-YEAR FORECAST

Before going into the details of the Tauri Group's forecast, it's worth reviewing the approach they used. The group used primary research and open-source materials to assess sRLV capabilities and also reviewed government budgets to build as complete and objective a picture of sRLV market dynamics as possible. Primary research used three techniques. The group interviewed 120 potential sRLV users and experts, including scientists and researchers, filmmakers, investors, educators, astronauts, physicians, consultants, private-sector researchers and spaceflight providers. They also surveyed more than 200 individuals with at least US$5 million in investable assets, in a randomized, scientific analysis, to estimate demand for sRLV flights among customers with assets consistent with sRLV prices. Finally, 60 suborbital researchers were polled at the 2012 Next Generation Suborbital Researcher's Conference (NSRC) and a space researcher's conference in Japan. Based on these data, the group analyzed each market and submarkets in terms of current and future activity, evaluated the sRLV market position based on sRLV capabilities, and compared this to competing services and platforms. The approach they took is summarized in Figure 2.4.

Each submarket was characterized in units relevant to the market segment. For example, locker equivalents were based on the standard sizes of Shuttle middeck lockers, EXPRESS (EXpedite the PRocessing of Experiments to Space Station) lockers, and NanoRacks (Figure 2.5) and, except for the crew, passenger seats were standardized by the number of individuals and not by their mass or volume.

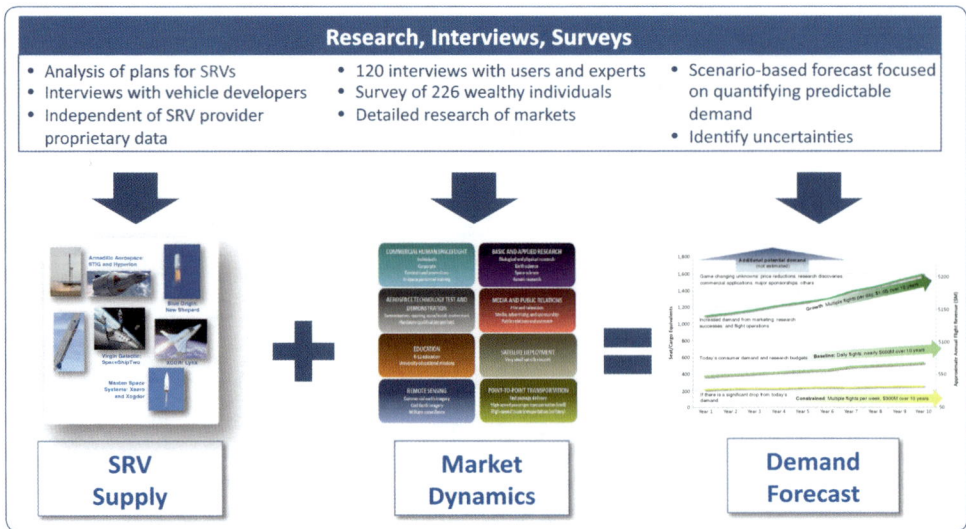

2.4   Tauri Group's research strategy. Courtesy: Tauri Group

Each market was then aggregated into seat/cargo equivalents which were determined using average cargo capacity compared to seats on proposed sRLVs. This aggregation resulted in the conversion of one seat/cargo equivalent equaling one seat or $3^1/3$ lockers – a translation that allowed the group to standardize and consolidate forecasts across all markets, reflecting a mix of cargo, people, and dedicated flights. The scenarios described earlier were based on the assumption sRLV prices would remain at current levels and sRLVs would be operated safely. Based on the operators who had announced seat prices, an average ticket price was estimated at US$123,000.

REUSABLE SUBORBITAL LAUNCH VEHICLES

We'll talk more about sRLVs in Chapters 4 and 5 but, for the purposes of understanding the survey, it's useful to have a brief overview of what these vehicles are and how they are operated.

sRLVs are launched beyond the threshold of space which, according to the FAI,[1] is 100 kilometers. During their brief excursion into space, these vehicles offer up to five minutes

---

[1]   The International Aeronautical Federation (FAI) is the world governing body for aeronautics and astronautics records, which includes man-carrying spacecraft. Among the FAI's responsibilities is the verification of record-breaking flights. Some records are claimed even though the achievements fail to meet FAI standards. For example, Yuri Gagarin earned recognition for the first manned spaceflight, despite failing to meet FAI requirements because he didn't land in his spacecraft (he ejected from it).

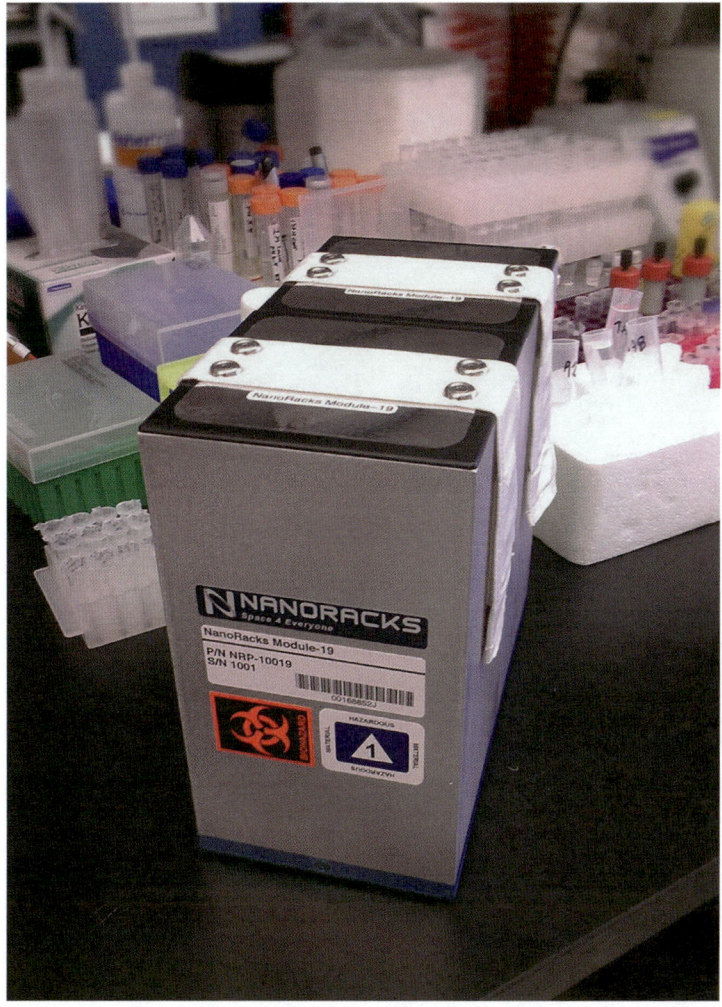

2.5 The NanoRack is an experimental platform built by NanoRacks LLC. Two NanoRacks are installed on board the International Space Station (ISS), each of the platforms holding up to 16 NanoRack payloads in the form of a CubeSat. NanoRacks LLC call the payloads NanoLabs, others call them CubeLabs™. Each payload is a compact 10 centimeters by 10 centimeters by 10 centimeters. In addition to being compact, the platforms allow for an easy "plug and play" interface, in which every project on the platform plugs into the ISS power and communications system; a similar approach could be used on board suborbital vehicles. Another popular cargo will probably be CubeSat-sized payloads. A CubeSat is a miniaturized satellite that has a volume of one liter (10 centimeters cubed) weighing no more than 1.33 kilograms, and usually uses commercial off-the-shelf (COTS) components. Thanks to their small size, CubeSats can be manufactured and launched for about US$65,000 to US$80,000 – a price tag far lower than most satellite launches, which is why it's a favorite option for schools and universities. Courtesy: NanoRacks LLC

2.6 SpaceShipTwo glides to a landing in the Mojave. Courtesy: The SpaceShip Company/
Virgin Galactic

of microgravity before returning to Earth. Some vehicles launch and land vertically, others
are slung under mother-ships and launched in mid-air, and some take off and land like
regular aircraft (Figure 2.6).

As you can see in Table 2.1, there are six companies in the business of planning, developing, and operating sRLVs, which have capacities ranging from tens to hundreds of kilograms. The vehicles, some of which can carry up to six passengers, have been in
development for years and, as with so many aerospace ventures, development timelines
have slipped and flight dates have been delayed. As this book is being written, George
Whitesides predicted Virgin Galactic revenue flights would start in 2014. Many in the
industry hope he's right. With so much uncertainty surrounding revenue flights, it isn't
surprising operators have been reluctant to provide details on how rapidly they will
increase flight rates, although most vehicle operators have targeted operational rates
between once per week to multiple flights per day.

MARKET ANALYSIS

Before looking at each of the markets identified by the Tauri Group, it's useful to understand how each was defined. We'll start with the Commercial Human Spaceflight market,
the expected market driver, which is divided into the following submarkets:

Individuals: space tourism and science/research flights
Corporate: flights as promotions for corporate customers
Contests and promotions: providing seats as a prize
Personnel training: training for orbital flights

Next is the Basic and Applied Research market, which is focused mainly on activities
that expand the pool of knowledge (this is different from the Education market, which

focuses on activities that use flights as a learning tool). This market is also divided into four submarkets:

Biological and physical research: payloads to investigate biological/physical responses
Earth science: observations and measurements of Earth
Space science: observations and measurements of the space environment
Human research: investigating human psycho-physiological responses

The Aerospace Technology Test and Demonstration market will advance technology maturity or achieve space demonstration, qualification, or certification. This market has two submarkets:

Demonstrations requiring space/launch environment
Hardware qualification and test

It is inevitable companies will use sRLVs to promote products and increase brand awareness. For example, the first job Astronauts for Hire (A4H) was hired for was to fly beer during a parabolic flight to promote an Australian beer company. The flight gained A4H a lot of media attention and no doubt the Media and Public Relations market will bring in a significant amount of revenue via the following four submarkets:

Film and television: filming for space-themed entertainment
Media advertising and sponsorship: logos/advertisements placed on space hardware
Public relations and outreach: awareness/recognition through association with SRLVs
Space novelties and memorabilia: objects that have flown in space

The fifth market is the Education market which – hopefully – will increase access to and awareness of space to schools, colleges, and universities. This will be achieved via the following two submarkets:

Education: payloads and activities for schools or students
University education missions: payloads developed by university students

Thanks to its payload adaptor, sRLVs such as the Lynx will be able to deploy satellites and launch small payloads into orbit. Depending on the success of XCOR, other operators may follow their lead and this (sixth) market – Satellite Deployment – could increase significantly.

The seventh market – Remote Sensing – includes the imaging of Earth and Earth's systems for commercial, civil government, or military applications. Finally, the eighth market – the Point-to-Point Transportation market (Figure 2.7) – will transport cargo and/or humans between distant locations, with three potential submarkets:

Fast package delivery (e.g. FedEx)
High-speed passenger transportation
High-speed troop transportation[2]

---

[2] Small Unit Space Transport and Insertion (SUSTAIN) is a concept first proposed in 2002 by the US Marine Corps to deploy Marines via suborbital flight to any location on Earth.

2.7 Reaction Rocket Engines' revolutionary transportation system. Courtesy: Reaction Engines

THE MARKETS

In their report, the Tauri Group analyzed each market in terms of market dynamics, how sRLVs fit into the market, current sRLV activity, sRLV demand forecast methodology, estimate of 10-year demand, uncertainty, and lack of awareness. What follows is a synopsized version of that report.

Commercial Human Spaceflight

This is the market that gets the lion's share of media attention and is divided into four submarkets:

- individuals;
- firms using seats as incentives or rewards;
- organizations offering seats in contests and promotions; and
- space agencies using SRLVs for training.

To date, just over 500 people have flown in space and only a handful of these flew commercial. In fact, since 2001, just seven leisure travelers (Figure 2.8) have purchased eight orbital flights (one passenger flew twice) for up to US$35 million per flight.

2.8 Richard Garriott. Courtesy: Richard Garriott

For those who don't have deep pockets, there is a whole industry offering space-related experiences that deliver key elements of the space experience, such as a view of the curvature of Earth, zero-G flights, and MiG fighter jet flights. Then there is the experience of training for spaceflight offered by the National Aerospace Training and Research (NASTAR) Center which has trained over 100 future tourists.

To wealthy individuals or to those willing to remortgage their house, suborbital flights are attractive because they offer a combination of space experiences combined with the bragging rights opportunity to say they're an astronaut, all at a price significantly lower than orbital flights. An added bonus is the training (three days in most cases) is much less than the six months required to train for an orbital flight. Plus, you don't have to learn Russian!

Thanks to its cachet of offering a trip into space, it won't be surprising that corporations may be one of the suborbital industry's best customers. After all, executives of large companies often receive incentives comparably priced to a suborbital seat, including items such as zero-G flights (the cost to charter a zero-G flight – for up to 36 participants – is US$165,000 or about US$4,600 per employee). Then again, while offering suborbital flights may be attractive as a bonus, those executives may be prohibited from flying under corporate insurance coverage for senior executives. Too bad.

Another source of income from flying passengers will come from contests. When I attended the NSRC conference in San Francisco in February 2012, XCOR offered the prize of a free flight on board their Lynx via a draw. Unfortunately I didn't win, but Jennifer Brisco did – she was presented with the award in May 2012 after the main winner couldn't accept the prize:

> "It's been my lifetime goal to take a suborbital flight. This is an absolute dream …
> I am in shock right now, I am speechless."
>
> Jennifer Brisco after being presented with the award at the Expo

A year later, XCOR hit the news again when UK-based Unilever bought 22 flights on board the company's Lynx as part of Unilever's space-themed AXE (brand of men's cologne) Apollo™ Campaign. The campaign included Apollo astronaut Buzz Aldrin and a 30-second Super Bowl ad.

Another popular use of sRLVs will be in-space personnel training for orbital activities because sRLVs offer long-duration microgravity, frequent launches, and opportunities to train for challenging physical or medical situations. For example, in June 2012, Excalibur Almaz, a company developing an orbital commercial human spaceflight vehicle, announced it will train its crews on XCOR's Lynx as a requirement for pre-mission training.

So where does the suborbital passenger-capable launch industry stand as we begin 2014? Well, there are a number of manned sRLVs in development, operated by companies that have booked 925 reservations, with ticket prices ranging from US$95,000 to US$250,000. The majority of reservations are for individuals, but a number of research flights have also been bought. Confirmed ticket holders include celebrities such as Ashton Kutcher, Tom Hanks, Brad Pitt (scheduled for the second flight), X-men director Bryan Singer, Formula 1 racing legend Michael Schumacher, Paris Hilton, and, of course, Sir Richard Branson himself. According to recent announcements by Virgin Galactic, 35–40% of deposits originate from the US, 15% from the UK, and 15% from the Asia-Pacific region. Incidentally, if you're a frequent flyer and would like to use those miles for a suborbital flight, it will cost you two million miles to redeem via Virgin!

Profiles of Select Suborbital Celebrities

Bryan Singer, 41, film and television director/producer
You know him from such movies as *The Usual Suspects* and *X-Men*. Singer, a science-fiction fan, who says *From the Earth to the Moon* is his favorite miniseries, met Sir Richard Branson at a hotel in Australia, where Branson described his plans to offer commercial spaceflights. Singer signed up.

Edward Roski Jr, 68, real estate developer, sports team co-owner
Roski has trekked to Mount Everest base camp, biked across Mongolia, and gone scuba-diving in New Guinea. In 2000, he chartered a submersible to tour the *Titanic*. Roski, who is co-owner of the Los Angeles Kings and the Los Angeles Lakers, figured if he had gone down that far, it would be nice to go up on the other side to see what Earth looked like from up there. He snagged ticket #128.

Victoria Principal, 57, actress
Best known for her role as Pamela Ewing on the 1980s television show *Dallas*, Principal signed on within the first 24 hours of Virgin's announcement. Another thrill-seeker who enjoys paragliding, bobsledding, and car racing, the Dallas star is so enthusiastic about the prospect of visiting space that she offered to join a test flight.

James Lovelock, 87, atmospheric scientist
More than 40 years ago, Lovelock worked at the Jet Propulsion Laboratory (JPL) where he marveled at images of Earth and Mars transmitted by satellites. The British scientist is best known for proposing Gaia theory, which suggests Earth is a living, self-regulating organism whose parts work together to sustain life. When he received a letter from Branson inviting him to go on a suborbital flight, he didn't hesitate.

Since there isn't an inexhaustible supply of celebrities, the Tauri Group was interested in forecasting demand among other groups with the financial wherewithal to afford a suborbital jaunt. Their assessment of demand for individuals included estimates among high-net-worth individuals (worth over US$5 million) and (poorer) space enthusiasts. To estimate demand for suborbital flights among high-net-worth individuals, the group conducted *The Tauri Group 2012 Survey of High Net Worth Individuals*, which revealed a relatively robust market of those willing to purchase suborbital flights. Analysis suggests there are enough – about 8,000 high-net-worth individuals across the planet – customers willing to pay current prices to constitute a sustained demand for suborbital flight. But it won't just be rich people flying. Although there are few reliable data available to predict the purchasing behavior of space enthusiasts, there will be some individuals with lower net worth who will spend a large proportion of their assets to purchase an sRLV flight. How many, we don't know, but several sRLV providers feel that more individuals outside the US$5 million population than predicted by the Tauri Group will seek to fly at current prices. Take Lina Borozdina-Birch for example:

Lina Borozdina-Birch, 38, chemist
Lina Borozdina-Birch says she has had two dreams since she was a girl in the former Soviet Union; one was to visit Disneyland and the other to visit space. In 1991, Borozdina-Birch came to the US and sought asylum. It wasn't long before she visited Disneyland and then, in 2004, the opportunity to realize her second dream came about following the launch of SpaceShipOne. Her husband, Jo, contacted Virgin and, after some deliberation, the couple took out a second mortgage on their home so that Borozdina-Birch could buy her ticket to space.

How many space enthusiasts like Borozdina-Birch and how many affluent individuals will fly will depend largely on the myriad factors that influence the market? In the best case growth scenario, potential customers' interest in suborbital flight will grow thanks to

Table 2.4 Forecasts for the Commercial Human Spaceflight Market.

Scenario	Year									
	1	2	3	4	5	6	7	8	9	10
Baseline	340	344	353	359	366	372	379	385	392	399
Growth	1,046	1,060	1,079	1,099	1,118	1,138	1,159	1,179	1,200	1,222
Constrained	187	188	191	195	198	202	205	209	213	216

increased marketing, publicity surrounding the start of human flights, and positive flight experiences. But, if the economy tanks, demand will fail. In short, if everything goes really, *really* well, the total number of seats purchased across all human spaceflight submarkets over the 10-year forecast period will be around 11,300, whereas if everything goes bad, the number of seats purchased across all submarkets totals just over 2,000 (Table 2.4).

The numbers in Table 2.4 look healthy, but you have to bear in mind there are many uncertainties and those numbers are based on assumptions. For one thing, it is impossible to predict the dynamics of demand as it responds to future events; demand may – and probably will – evolve in unpredictable ways. For example, demand may grow more rapidly than predicted based on "me too" effects, following exciting launch experiences. Equally, demand could decline if a large proportion of individuals report unpleasant flight experiences such as space motion sickness. Also, the forecast assumes individual passengers fly once only, that only 40% of interested passengers will fly within the next 10 years, and that most passengers have net assets exceeding US$5 million; relaxing any of these assumptions will increase demand significantly. For example, if 80% of interested passengers fly in the next 10 years, the forecast doubles!

Basic and Applied Research

There are some who argue the main source of revenue for the commercial suborbital industry will come not from rich tourists, but from research institutions and universities paying for science flights. In fact, the premise upon which A4H was created was in anticipation of such science flights and the expectation that organizations will hire a cadre of highly trained astronauts to conduct science experiments. It wasn't long before operators also realized a sizeable slice of their revenue will likely come from science flights and set about reconfiguring their cabins.

With the Lynx and SpaceShipTwo flying, sRLVs will support many types of space-related research, which are generally grouped into four disciplines: space science, biological and physical research, Earth science, and human research. The sRLV capabilities most useful for research are access to the space environment, access to microgravity (for biological and physical research), transit through the upper atmosphere (Earth science), and access to passengers being subjected to acceleration and deceleration.

There may be some wondering just how popular this market will be. After all, isn't most space research being conducted on board the ISS? Well, yes it is, and that is precisely why suborbital research flights will most likely be *the* revenue driver of the industry, the reason being the protracted and costly process of having an experiment accepted on board the ISS – a process that can take years and years. Science on board sRLV flights on the other hand will be a much more stream-lined process (see Chapter 8).

There are four research applications – *microgravity research, atmospheric science, suborbital astronomy*, and *human longitudinal research* – particularly suited to sRLV capabilities and these support a wide range of experiments. Let's take the first of these – microgravity research.

There are a number of niche applications for microgravity research including experiments that can use the five-minute sRLV microgravity window – experiments such as the crystallization of particles in a charged plasma that can't be modeled by computer simulation, experiments that don't have adequate terrestrial research alternatives, and those requiring human tending. A bonus for scientists flying these experiments is that sRLVs provide frequent research opportunities at a lower cost and these vehicles are more accessible than orbital systems, albeit for a shorter microgravity duration.

The second research application – Atmospheric Science – is particularly suited to suborbital flight profiles because they offer scientists the opportunity to study the upper atmosphere (sounding rockets can also be used but at a high cost). And, given the flexibility of suborbital flight profiles and the diverse launch locations, trajectories can be tailored to visit all sorts of regions of interest.

Suborbital Astronomy – until the advent of sRLVs, the opportunity for high-quality astronomical observations could only be conducted using orbital telescopes, which always have very (very!) long waiting lists. But sRLVs reach altitudes that provide access to ultraviolet (UV) and infrared (IR) spectra, which contain useful astronomical information. And, budget sRLV flights will provide an opportunity for launching low-cost telescope payloads that may observe phenomena deemed too risky for billion-dollar orbital telescopes.

The fourth of these applications – Human Longitudinal Research – will provide a mother-lode of physiological data for space life scientists because the data sets of humans flying suborbital flights can be counted on the fingers of one hand. Researchers will want to study the physiological responses to microgravity and especially the acceleration transitions that occur during ascent and descent. They will also want to understand the mechanical responses in the vascular system, cell structure, and chemical changes in immune pathways, because these responses are poorly understood in suborbital flight. Then there is the issue of pharmaceuticals that minimize the discomforts (space motion sickness) some passengers may experience. While conducting human research on scientist astronauts won't be a problem, the extent to which suborbital passengers will wish to participate has yet to be determined, although the indications are that some are willing to be guinea-pigs.

Given human spaceflight will represent the biggest slice of the market, it's likely the best-funded research will be in the field of human research. Funded by such entities as the National Space Biomedical Research Institute (NSBRI), a non-profit science institute established by NASA in 1997 to research safe human spaceflight, this research will start as soon as revenue flights start, and will likely grow to a large clinical research trial by the end of the forecast.

As far as non-commercial microgravity research is concerned, the Tauri Group estimate this will be driven by internal research funding from universities, and augmented by non-profit organizations. In its forecast, university spending on sRLV microgravity experiments starts at about US$200,000 and grows to over US$500,000 in the last year of the forecast. It doesn't sound like a lot of money, but the problem with microgravity research is there is no well-understood commercial application for this type of research on sRLVs – at least not among the biotech, pharmaceutical companies, and technology-focused venture capital firms analyzed as part of the study. However, some venture-focused companies predicted there would be some level of exploratory commercial research, to enable firms to gain insight into sRLV capabilities and assess how those capabilities might benefit their research portfolios. Investing in research projects without connection to clear commercial outcomes sounds like a non-starter, but there is always the *what if* factor that may drive exploratory research, and it is this the Tauri Group think will start slowly and increase to a total of about US$5 million annually.

In terms of *where* this money will be spent, it is anticipated that, in the early days, the US government will be the primary user of sRLVs for basic and applied research, funding astronomy, atmospheric research, and longitudinal human research. But, after five or six years of revenue operations, other nations' governments will begin to implement larger standing sRLV programs; the Tauri Group suggest as many as 50 nations may be potential sRLV users for government-funded research, based on current research activities and budgets.

Aerospace Technology Test and Demonstration

Another use of sRLVs will be to advance technology maturity or achieve space demonstration, qualification, and/or certification. In the Aerospace Technology Test and Demonstration market, payloads will be tested and/or demonstrated on sRLVs to qualify and/or obtain data on flight systems in development. These payloads may be at any level of maturity, but will most likely be at the higher technology readiness levels (TRLs) that require test or demonstration in relevant environments (Table 2.5).

Today, space agencies and companies conduct test and demonstration activities in terrestrial facilities and during spaceflight. Terrestrial facilities include rocket test stands, thermal chambers, vacuum chambers, drop towers, and wind tunnels. In addition to using these facilities, computer modeling is being used more and more, thereby reducing requirements for high-fidelity tests. Spaceflight test and demonstration activities use sounding rockets such as the Black Brant (Figure 2.9), the ISS, and other platforms such as evolved expendable launch vehicles (EELVs), although this latter category is predominantly associated with missile defense.

Given that a typical sounding rocket test can cost anywhere between US$2 and US$5 million or more, sRLVs are likely to be very popular supporting test and demonstration activities, especially since sRLVs have the potential for human interaction. Like sounding rockets, sRLVs provide access to high altitudes, upper atmosphere aerodynamics, and can provide access to similar thermal, radiation, and vacuum environments as orbital space systems, although sRLV performance in these environments is mitigated by just five minutes of exposure. How much business sRLVs will take away from traditional platforms is

Table 2.5 Technology Readiness Levels

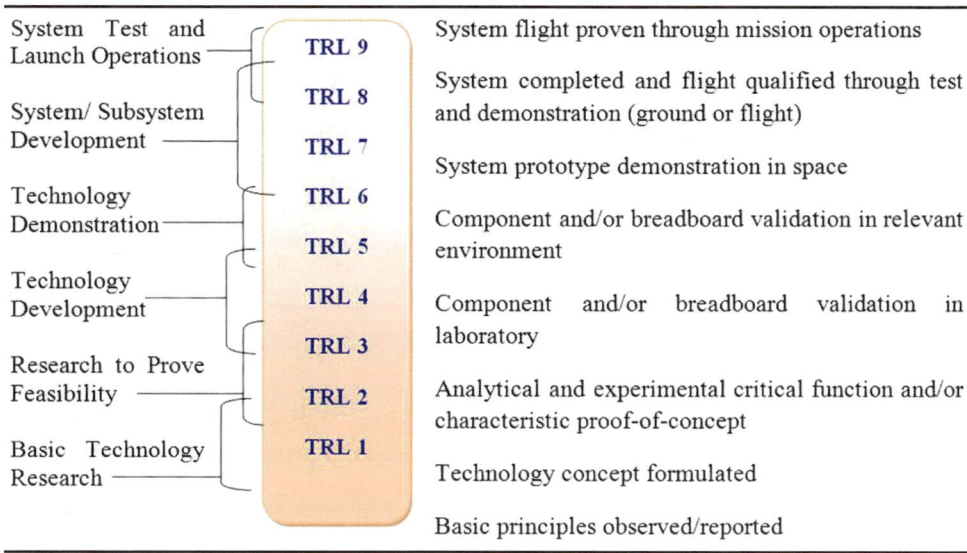

System Test and Launch Operations	TRL 9	System flight proven through mission operations
	TRL 8	System completed and flight qualified through test and demonstration (ground or flight)
System/ Subsystem Development	TRL 7	System prototype demonstration in space
Technology Demonstration	TRL 6	
	TRL 5	Component and/or breadboard validation in relevant environment
Technology Development	TRL 4	Component and/or breadboard validation in laboratory
	TRL 3	
Research to Prove Feasibility	TRL 2	Analytical and experimental critical function and/or characteristic proof-of-concept
Basic Technology Research	TRL 1	Technology concept formulated
		Basic principles observed/reported

hard to say, but the Tauri Group estimate up to 25% of orbital test and demonstration missions require capabilities that could be provided by sRLVs. If you're interested in testing a payload on board an sRLV, you can apply through NASA's Flight Opportunities Program, which supports technology payloads on sRLV precursors and alternatives, like parabolic platforms; through this program, NASA has already procured payload capacity from Virgin Galactic. In fact, NASA and other space agencies will likely be the major users of sRLVs for test and demonstration in the near future, with half of the sRLV-suitable payloads previously launched on the Shuttle estimated to transition to sRLVs (the other half are better matched to the capabilities provided by the ISS).

Media and Public Relations

On April 18th, 2013, XCOR Aerospace General Sales Agent (GSA) Space Expedition Corporation (SXC) CEO Michiel Mol announced the latest Lynx Mark I flight participant would be Japanese film actor Koichi Iwaki. Mr Iwaki, who also happens to be a spokesman for the US watch company Luminox, an SXC affiliated partner, has starred in over 50 films and TV series since 1975. Andrew Nelson, XCOR's Chief Operating Officer, noted:

> "We are thrilled to announce that SXC has welcomed an exceptional artist – both on the screen and the road – to the growing list of Lynx flight clients. Iwaki-san is known as a groundbreaking artist in Japan and around the globe, and will be the first Japanese citizen to fly on board a Lynx and I suspect he'll be wearing his Luminox space watch."

2.9   Black Brant sounding rocket. A sounding rocket basically comprises solid-fueled rocket motor(s) and a payload. The Black Brant has a two-stage solid-fueled rocket; the first stage is a Terrier booster and the second stage is a Black Brant motor. Courtesy: NASA

Just one example of how sRLVs may shape the Media and Public Relations market! Incidentally, if you would like to follow in Mr. Iwaki's rocket plume, you can buy a flight on the Lynx by going to spacexc.com or xcor.com.

This market includes activities that use spaceflight to promote products, increase brand awareness, and film what happens during a flight. Of these submarkets, perhaps the potentially most lucrative is film and television for the simple reason people love space-themed movies, as evidenced by the fact that about 5% of all major feature films have been space-related and, of the top 50 domestic grossing movies of all time, eight are space-related. Films and television programs actually filmed in space are less common, although *Apollo 13* director, Ron Howard, filmed 3 hours and 54 minutes of raw footage over 612 parabolas to simulate the in-space scenes.

The *Apollo 13* Space Scenes

When it comes to making believable space movies, the special effects gurus can only take you so far. Sure, they have ways of moving the camera, they can suspend astronauts in harnesses and even "float" objects on invisible threads, but when it comes to replicating weightlessness, these tricks don't cut it. In fact, even the most famed directors get it wrong. Take the iconic *2001: A Space Odyssey*, which features the most famous piece of weightless fakery in film history. In the early part of the film, a space traveler sucks food up a straw, only to have the food slide back down the straw when he takes it from his lips; in real zero-G, that wouldn't happen.

So, when Tom Hanks, Bill Paxton, and Kevin Bacon signed on to make *Apollo 13*, which tells the story of how NASA rescued astronauts Jim Lovell, Fred Haise, and Jack Swigert on the ill-fated third Moon-landing mission, director Ron Howard and his cast paid a visit to Bob Williams, NASA's test director for the Zero-G Aircraft Program. Howard was so impressed with the experience that he began thinking about how to shoot the weightless sequences inside the aircraft by building a mock-up of the interior of the Apollo capsule inside the Vomit Comet – which is exactly what Howard did. He and his crew broke down each weightless scene into segments shorter than 25 seconds, and had the actors act while being bounced around inside the plane.

It made for a tough shooting schedule that included nearly four hours of weightlessness, which is more than most astronauts get before their first flight. To begin with, efforts were frustrating, with actors drifting in and out of shots, crewmembers landing on hard edges, and generally a mad scramble. Then there was the sound of the KC-135 engines, which screamed so loudly that none of the dialogue was usable. But, with practice, the cast and crew adapted, making two flights a day with 30–40 parabolas on each flight. The end result was convincing – even to real astronauts. In fact, when some of the Apollo astronauts were invited to a sneak preview, the theater exploded with applause when the movie ended.

The use of suborbital flights to film scenes for a movie could prove appealing if the scenes are of relatively short duration, because it would be expensive, not to say logistically challenging (at least one provider has received a proposal to produce adult films during flight, which was declined), to have to do what Ron Howard did when making *Apollo 13*. Commercials on the other hand are more suitable, and could prove popular with companies looking for an edge to promote their product. Flight rates could also be boosted by the placement of logos and advertisements on the vehicle (echoing the placing of the Kodak logo on the outside of the ISS in 2001) and its occupants. Some companies will also pay money to have their product flown in space. Take A4H, for example. Shortly after I joined A4H, the organization received its first job offer to fly an Australian beer in space. In case you're wondering what NASA's position is on this, the agency forbids the consumption of alcohol. But, with space tourism just around the corner, the idea of testing beer in space seemed well timed to take advantage of the impending market. And, since research would be performed on the effects of alcohol on astronauts in a microgravity environment, the "beer in space" flight would have scientific value! The deal was signed between A4H, the 4-Pines Brewing Company, and Saber Astronautics Australia. Rather than studying the physics and chemistry of carbonation or fermentation, the research focused on the human experience of consuming the beverage in microgravity. The A4H flight researcher – Todd Romberger (who happened to be a home brewer) – sampled the beer during weightless parabolas on a series of Zero-G Corporation flights and recorded body temperature, heart rate, blood alcohol content, as well as reporting on the beverage's taste and drinkability.

For those who are interested in space beer, the box for the terrestrial six-pack version of the Vostok Space Beer read:

"From the dawn of civilization, people have brewed beer and wherever people went, beer followed. In the Middle Ages, beer helped monks survive long periods of fasting. Sailors drank beer (with a splash of lime) to stave off scurvy. During the age of exploration, distances traveled necessitated brewers to get creative, making beer recipes to last the voyages that would bring this great beverage to the world!

To continue this great tradition we have created a beer for another momentous voyage. A beer specifically designed for the next frontier of space exploration. Space engineers from Saber Astronautics Australia have teamed with 4 Pines in our own space race; to take this 4 Pines stout which tastes fantastic on Earth and adapt it for space to produce the world's first space beer.

Enjoy! (no matter where on Earth or in space you are)"

As for Todd, this is what he had to say:

"It was exciting to participate in this ground breaking research. I'm happy to say the flight was a success. We provided the customer with the data they needed, and made a little history along the way. Not to mention, the beer was delicious!"

Also included in this market are novelties and memorabilia. Take Celestis Memorial Spaceflights. Celestis is a service that will fly cremated remains into suborbital space, Earth orbit, onto the lunar surface, or even into deep space. To date, the company has flown its canisters on 11 launches, carrying the remains of over 800 individuals. Other memorabilia companies include Space Wed, a wedding ring company, which flew 50 sets of wedding rings into space in May 2011, and To Space, which has brokered sending custom payloads into space, such as business cards and personal items. Even commercial spaceflight juggernaut, Bigelow Aerospace, got in on the action by flying small personal items for US$300 while testing the Genesis II module (the items were visible via a webcam on the company's website).

Tauri's forecasting for this market was based on costs, historical data, and interviews, which indicated interest in producing films and TV programming on sRLVs will be related to the novelty of the experience and the association with suborbital spaceflight. It's a difficult market to forecast because some suborbital companies will allow sponsorships, advertisements, and commercials on their vehicles, while others will probably limit third-party sponsorship due to brand maintenance. Then there is the difficulty predicting the public response to sRLV marketing, flight activity, and the reports of flight experience; if several passengers return complaining about motion sickness, then this market will suffer. What will likely happen is there will be a near-term surge in interest that fades as suborbital flights become more common. Let's hope not, but with attention spans reducing by the week, it's possible.

Education

This should prove to be a dynamic market because sRLVs provide opportunities for schools, colleges, and universities to increase access to and awareness of space, especially through the flight of student-built payloads. Then there is the potential of teacher-in-space and student-in-space programs, so it's difficult to see how this market won't drive up the flight rate. In 2013, students who want to fly their experiments in space could take advantage of the CubeSat program, which can cost in the tens of thousands of dollars, although universities can sometimes use government-sponsored complimentary rides to orbit as secondary payloads. With more affordable suborbital flights on the horizon, there are several projects already scheduled to be flown on the new fleet of sRLVs as part of the research and education mission (REM) program. And the REM market isn't limited to the US; at a price of US$250,000 per seat, virtually every one of the 190-plus nations on Earth can afford an education mission. Think of the returns – motivating students into science careers by seeing physics and chemistry classes, and doctoral theses streamed from space on a daily basis. No doubt about it, education missions will be a game-changer for the simple reason that sRLVs can potentially enable novel and unprecedented levels of participation by students in space. And, with such frequent launches, schools won't have a problem aligning projects with academic calendars. The only hurdle operators will have is developing awareness of the possibilities of sRLV programs among science teachers because this will be a marketing effort that may come with a high marketing cost.

Already, there are several ongoing efforts to develop this market. In Florida, an effort is underway to develop integrated curriculum modules that use sRLV services and link those activities to state curriculum requirements. The Space Frontier Foundation, a space industry advocacy organization, is sponsoring a non-profit project called "Teachers in Space" that will sponsor teachers to fly on sRLVs. In fact, XCOR has donated three seats to this program. Then there is UP Aerospace, which has launched educational payloads (funded by NASA or New Mexico state funds) through the New Mexico Space Grant Consortium that includes 29 public high schools and 17 universities. The Tauri Group expects this market to be dynamic, forecasting three teacher seats in the baseline forecast, growing to 60 total seats over the 10-year forecast period, which is similar in expenditures to a previous program that flew 1,300 educators on Zero-G.

Table 2.6 Satellite Categories.

FAA categories	Mass and non-standard classifications (kg)
Heavy	>9,072
Large	4,537–9,072
Intermediate	2,269–4,536
Medium	908–2,268
Small	92–907
Micro[1]	up to 91

[1]Includes "very small satellites" (<15kg), Nanosats (1–10kg), CubeSats (1–2kg), Picosats (0.1–0.9kg), and Femtosats (<0.1kg).

Satellite Deployment

sRLV satellite deployment is the launch of satellites weighing under 15 kilograms. These micro-satellites, which can include single application satellites or large constellations of redundant satellites working together, are mated to a propulsion stage and launched into low Earth orbit (LEO). The very small satellite (Table 2.6) market gained traction with the introduction of the CubeSat (a 1-kilogram, 10-centimeter cube) and the Poly-Picosatellite Orbital Deployer (P-POD) deployment system, which allows for rapid payload development and standard launch interfaces. Between 2002 and 2011, 105 satellites under 15 kilograms were launched, primarily for universities in the US. It's a market that is continuing to grow, especially among civil and defense agencies, which are using more and more very small satellites and developing new capabilities and supporting infrastructure. For example, the US Army is developing the Kestrel Eye imaging satellites and SNAP communication satellites (weighing under 25 kilograms) and the National Reconnaissance Office (NRO) has a CubeSat program and has already purchased 10–20 CubeSats, with an option to purchase up to 50.

Most very small satellites have been deployed as piggyback payloads on low-cost, commercial launch vehicles, primarily Dnepr, while larger payloads launch from a custom EELV using adapters that allow for several satellites to be launched together. There are more than a dozen EELVs sized and designed specifically to deploy very small satellites such as the US Army's Soldier Warfighter Operationally Responsive Deployer for Space (SWORDS), a nanosatellite launch vehicle, and the Defense Advanced Research Agency (DARPA)'s Airborne Launch Assist Space Access (ALASA) program. As commercial suborbital spaceflight moved closer to reality, it wasn't surprising to see awards to Virgin Galactic, which leverages the WhiteKnight2 carrier aircraft for an orbital launch system. Then there is XCOR, which has announced plans to serve the very small satellite deployment market, using its Lynx Mark III that will carry very small satellites (12 kilograms) to LEO from a dorsal pod on flights beginning in 2017.

The Tauri Group forecasts the demand for very small satellites will continue to increase through the end of the forecast with about 100 satellites of 15 kilograms and under being launched worldwide in Year 10 and 760 satellites – split evenly between civil, military, and university – launched over the 10-year forecast. As ever, the forecast assumes various factors, such as continued support for the NASA CubeSat Launch Initiative, which should energize the suborbital community, resulting in more payloads and providing more launches. The forecast also assumes the US defense community will be a potential customer for sRLV satellite deployments in the near term.

Remote Sensing

This market is the use of sRLVs for the acquisition of imagery of Earth for commercial, civil government, or military applications. As you can see from Table 2.7, this is already a robust market served by aerial and satellite platforms, which are well established. In fact, due to customer requirements such as timeliness, accuracy, resolution, quality, and other parameters, sRLVs might find it difficult to break into this market due to over-flight restrictions, high launch and re-entry velocities, and fixed runways.

Given the large increase in the use of UAVs for intelligence, surveillance, and reconnaissance in the past decade, it's difficult to see how sRLVs fit into this market, but they could create a niche between aerial and satellite remote sensing in terms of swath width, resolution, and revisit time. sRLVs also have potential applications in civil, commercial, and military remote sensing in the areas of disaster management, border security, policy enforcement, pipeline surveillance, and agriculture. Also, for military applications, sRLVs could ascend in friendly airspace and achieve views of hostile territory without violating airspace restrictions or exposing the vehicle to the threat of engagement. That said, this is a marginal market, and it is unlikely to drive flight rates appreciably.

Point-to-Point Transportation

Imagine a flight from London to Los Angeles and you'd probably think of being stuck in an airplane seat for the best part of a day. But, if XCOR has its way, the endurance test that is a transcontinental flight may have an expiry date because the company believes future commercial flights may take passengers to the edge of space to dramatically reduce travel time. So, instead of spending 13 hours flying from, say, New York to Tokyo, your edge-of-spaceflight time would be reduced to just 90 minutes. All thanks to the Lynx, a two-seat rocket plane which can utilize conventional airport runways for horizontal take-off and landing.

This future accelerated point-to-point (P2P) service between distant hubs is thanks to the fact that the Lynx can take advantage of many more take-off and landing options than spacecraft that require dedicated facilities. Not surprisingly, there is significant interest in P2P services, but it is unlikely to be realized before the turn of the decade because there are still several technologies that need to be matured – aerodynamics, hypersonic, guidance, navigation, and control, propulsion, high-temperature materials and thermal protection systems, and fuel/propellant storage, to name but a few. But, while we may have to wait a few years before taking that hypersonic trip across the pond, sRLVs will play a key role in setting the stage.

The promise of hypersonic flight sending us halfway around the world in a matter of hours may also be realized thanks to Reaction Rocket Engines, a British company that reckons its hypersonic engine will send us streaking across the sky at speeds well over Mach 5. The hypersonic engine design (Figure 2.10) includes a novel way of cooling the

Table 2.7 Remote Sensing Market.

Platform	Resolution	Swath width	Revisit time	Spectrum	Remarks
Aerial imagery (piloted)	Very high resolution (cm)	Tens of km^2	On demand	Any	Established market Flights regulated Over-flight permission required Moderate investment High cost per image
Aerial imagery (UAVs)	Very high resolution (cm)	Tens of km^2	On demand	Any	Emerging market Flights not formally regulated Over-flight permission required Low investment Moderate cost per image
SRLVs	High resolution (m)	Hundreds of km^2	On demand	Any (due to short loiter time, may not be ideal for radar)	Emerging market Launch/re-entry regulated (US) Over-flight permission required Moderate investment High cost per image Currently limited viewing area
Satellites	Low to high resolution (km to m)	Thousands of km^2	Days	Any	Established market Some regulation No over-flight permission required Shutter control High investment Low cost per image Revisit time can be enhanced by increasing number of satellites in one orbit

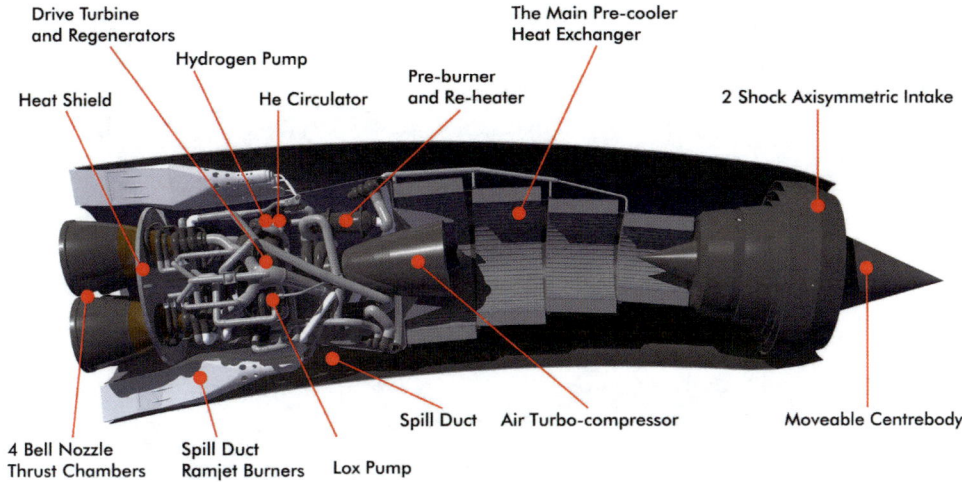

Drive Turbine and Regenerators · Hydrogen Pump · Heat Shield · He Circulator · Pre-burner and Re-heater · The Main Pre-cooler Heat Exchanger · 2 Shock Axisymmetric Intake · Spill Duct · Air Turbo-compressor · Moveable Centrebody · 4 Bell Nozzle Thrust Chambers · Spill Duct Ramjet Burners · Lox Pump

2.10   Reaction Rocket Engines' SABRE engine. Courtesy: Reaction Engines

air for an engine that will use oxygen in the atmosphere up to Mach 5.5 before switching to rocket power for the ride in space.

Until Reaction Rocket Engines came along, the big problem had been propulsion, because at speeds beyond Mach 2 or so, a jet engine has trouble getting the oxygen needed for combustion, although Kelly Johnson's SR-71 Blackbird design used creative ways to deal with the incoming air needed to achieve record-setting Mach 3+ speeds. But, beyond Mach 3, it gets really tough, mainly due to heat issues. Of course, rocket engines can achieve hypersonic flight, but the vehicles use multiple stages. To avoid carrying a supply of oxygen, as rockets do, some engineers figured they could develop an air-breathing design that could operate in the hypersonic speed range as a first stage. And now Reaction Engines reckons its cracked the problem – their secret is cooling the air as it enters their hypersonic SABRE engine, which uses pre-cooler technology to cool the incoming air-stream from over 1,000°C to −150°C in less than 1/100th of a second. If all goes well over the next several years, the air-breathing engine will accelerate a vehicle to about Mach 5.5, after which a liquid oxygen tank will supply a rocket engine for part of the flight in space. But unlike today's SS2, there will only be one stage involved for the entire flight thanks to the SABRE design. To date, Reaction Engines has completed more than 100 test runs of the cooling system and it hopes to have a subscale ground engine running by 2015. Sounds promising but, as the X-51 Waverider (Figure 2.11) team discovered, hypersonic flight is a difficult nut to crack.

But, if the hypersonic nut *is* cracked, there are still some technical, logistical, legal/regulatory, and economic barriers that have to be addressed. First of all, a P2P infrastructure needs spaceports at each destination, and these spaceports have to be integrated with other modes of transportation and local air traffic control systems. Then there are the issues of international air systems, over-flight restrictions, environmental regulations, insurance.

2.11 Waverider. Courtesy: NASA

Several years ago, there weren't many people in the industry sold on the idea of using sRLVs for anything but tourist flights. But, with NASA stating the agency is now open to flying people on suborbital vehicles, the thinking on the subject of suborbital research and other markets is much more of a reality:

> "We absolutely do not want to rule out paying for research that could be done by an individual spaceflight participant – a researcher or payload specialist – on these vehicles in the future."
>
> Lori Garver, NSRC, June 3rd, 2013

Thanks to this policy change, operators now have a backlog of flights that serve the markets discussed in this chapter. In fact, at NSRC 2013, NASA representatives of the agency's Flight Opportunities program appealed for more applications, saying there weren't enough being submitted. And, while many bemoan the delays in getting the vehicles flying, once routine flights begin, the potential identified by the Tauri Group may finally be realized.

3

Training Suborbital Astronauts

If you'd like to get a feel for how the new cadre of commercial suborbital astronauts will be trained, you can sign up for the National Aerospace Training and Research (NASTAR) Center's Basic Suborbital Space Training course, which will cost you US$3,000.00. Or you can contact any one of a number of training companies such as Suborbital Training[1] and have a customized suborbital training course designed just for you. Given that astronauts have been launched into space for more than five decades, you'd think the training would be figured out by now, but remember we're talking about *suborbital flight*[2] and *commercial astronauts*. Also, the operational experience for manned suborbital spaceflight is very limited, consisting of just two Mercury-Redstone rocket flights in 1961, two X-15 flights in 1963, a Soyuz launch abort in 1975, and three SpaceShipOne (SS1) flights in 2004. Also, if we applied the rigorous NASA medical certification and training standards to commercial astronauts, we'd find ourselves with a very small pool of candidates because not everyone has the résumé of a government astronaut. And, finally, the stresses of suborbital flight are by several orders of magnitude smaller than those imposed by orbital flight, so it stands to reason that training should reflect this. But how much training should a commercial suborbital astronaut have? Should this new cadre of astronauts complete the same training as spaceflight participants (space tourists) or should they be required to meet more rigorous standards? And what about the medical standards?

---

[1] Suborbital Training (www.suborbitaltraining.com) offers customized training that includes vehicle familiarity, anti-G straining instruction, high-altitude indoctrination, emergency egress training, zero-G exercises, and the theoretical/practical aspects of payload integration. The company also offers the services of a corporate astronaut.

[2] According to the Fédération Aéronautique Internationale (FAI), a suborbital flight has to reach an altitude higher than 100 kilometers above sea level, although the US Air Force (USAF) and the Federal Aviation Administration (FAA) consider 80 kilometers as the altitude to qualify as space flight. 100-kilometer altitude also happens to be the boundary between Earth's atmosphere and space – the Kármán Line, named after Theodore von Kármán, a Hungarian-American engineer and physicist who calculated that, at 100 kilometers, the atmosphere becomes too thin for aeronautical purposes because any vehicle would have to travel faster than orbital velocity to derive sufficient aerodynamic lift from the atmosphere to support itself.

E. Seedhouse, *Suborbital: Industry at the Edge of Space*, Springer Praxis Books,
DOI 10.1007/978-3-319-03485-0_3, © Springer International Publishing Switzerland 2014

MEDICAL STANDARDS

Let's start with medical standards but, before we do, it's useful to familiarize ourselves with the suborbital flight environment and the potential medical risks imposed by such a flight. Since many reading this have heard of Virgin Galactic's SpaceShipTwo (SS2), we'll use this vehicle's flight profile as a reference.

SS2 will carry two pilots and up to six spaceflight participants. The cabin will be pressurized to an altitude of 2,440 meters and passengers will breathe re-circulated atmospheric air. Flights begin with a horizontal take-off underneath the carrier aircraft WhiteKnightTwo (WK2). Once WK2 reaches an altitude of 15,000 meters, SS2 is launched. During the 70-second boost phase, passengers will be subject to acceleration loads up to 3.8 G and speeds up to Mach 3 (maximum speed will be 4,180 kilometers per hour), 30 seconds after the rocket fires. The suborbital coast phase will last about four minutes and SS2 will reach a maximum altitude of 110 kilometers. During the deceleration phase, passengers will experience up to 6 G, but the seats will recline to convert most of the forces to +Gx (we'll get to an explanation of G-forces[3] shortly). To increase stability and drag for re-entry, SS2's wings will rotate to a feather position. Then, at an altitude of 24,500 meters, the 25-minute glide phase will begin with a return to an unpowered horizontal runway landing. Total flight duration will be about two and a half hours. The risks? Well, a suborbital flight is not nearly as demanding as an orbital flight but there are still some risks which we'll cover here.

SUBORBITAL MEDICAL ENVIRONMENT

If you watched the SS1 flights, you don't need to be a flight surgeon to know these flights will expose you to an environment much riskier than what you experience when flying on a 767. So, if you happen to have any pre-existing medical conditions, it's probably worth seeing a physician just to make sure a suborbital flight won't make things worse. For the general population, most medical issues related to suborbital spaceflight are fairly straightforward because suborbital flyers don't have to worry about the more serious medical problems associated with orbital flight, such as bone loss, and radiation exposure. But there is still that G problem, especially the rapid change from acceleration launch forces to zero-G weightlessness followed quickly by re-entry deceleration. Even a relatively benign flight profile like SS2's is provocative enough to make even the most ardent roller-coaster fan a little queasy. And, medically, these transitions could lead to cardiovascular and neurovestibular effects that are currently undefined because there haven't been many who have flown suborbital flights. Although you can test your G-tolerance in a centrifuge (Figure 3.1) and your zero-G-tolerance in parabolic flight (Figure 3.2), it's impossible to

---

[3] A "G" is a measure of the acceleration of an object normalized by the acceleration caused by gravity. Without considering air resistance, gravitational pull causes free-falling objects to change their speeds by a constant of 9.81 meters per second^2. Dividing acceleration (calculated as change of velocity divided by time) by this constant yields the acceleration in G.

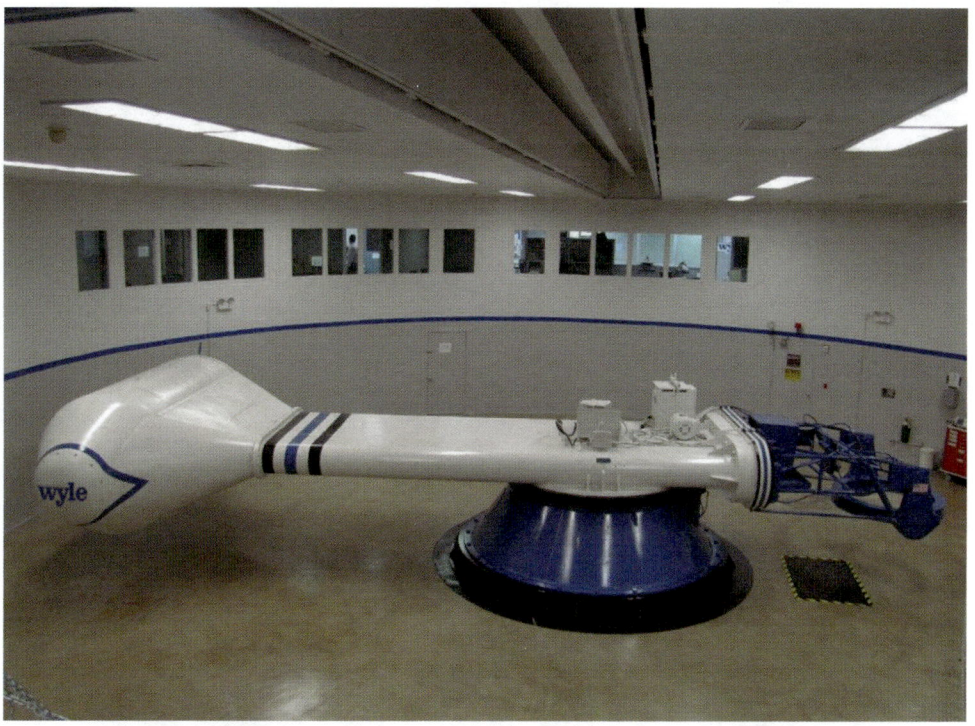

3.1 Human centrifuges such as this one at Wyle Labs will be used to train spaceflight participants. Courtesy: Wyle/NASA

3.2 ESA uses a modified Airbus for its parabolic flights. Courtesy: ESA

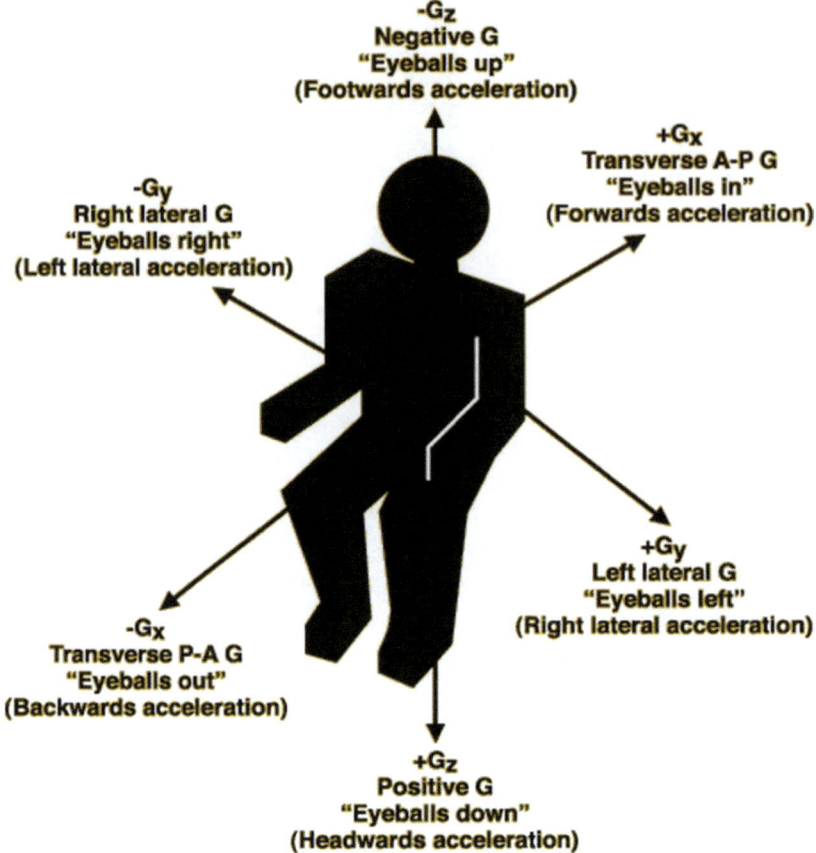

3.3 The different types of acceleration. Courtesy: NASA

simulate the forces of the suborbital flight environment except for the actual experience of suborbital spaceflight.

Acceleration

From the flight surgeon's perspective, some of the most troubling issues of a suborbital flight profile are the launch *acceleration* and re-entry *deceleration*, especially when the acceleration exposure is in the head-to-foot ("eyeballs down" or +Gz) direction (Figure 3.3). That's because Gz acceleration can cause a number of neurovestibular, cardiovascular, *and* musculoskeletal problems. Exposure to Gz can also affect pulmonary function proportionally to its applied force magnitude – for example, at the lower end of the G scale, say 2–3 G, most people will experience difficulty breathing, while at the other end of the G-load spectrum, say 5–6 G, there is a risk of airway closure.

Obviously, spacecraft designers want make your experience of riding a rocket as safe as it can be, so they try to limit launch and re-entry acceleration forces by ensuring most of the acceleration is in the +Gx ("eyeballs in") direction. That's because people are more tolerant to +Gx acceleration and, with the heart and brain located at approximately the same level within the acceleration field, there is less risk for *gravity-induced loss of consciousness* (G-LOC) or *almost loss of consciousness* (A-LOC). Acceleration stress is one of the issues that most worry flight surgeons because it is *dysrhythmogenic*, which means the heart's rate, rhythm, *and* conduction can be upset. In fact, high G-forces, and/or particularly long exposures to acceleration, could potentially increase the frequency of a heart problem known as a *dysrhythmia*, which is why spaceflight accelerations have for the most part been designed to be in the +Gx axis.

A G Primer

When you are exposed to an increase in +Gz (head-to-toe acceleration), the pressure required to perfuse your eyes and brain increases and blood begins to pool in the large blood vessels of your legs. As the G levels ramp up, the perfusion pressure requirements increase and the volume of blood returning to your heart decreases further. Making things worse, your eyes and the brain receive a decreasing amount of oxygenated blood. And, if the duration of the exposure is long enough, your eyes, which need a certain amount of perfusion pressure to function, will manifest symptoms such as a loss of peripheral vision, which can proceed to a total loss of vision if the acceleration level is high enough and long enough. Then, as the acceleration level and duration increases, you will lose consciousness – G-LOC – and only regain consciousness once the acceleration level is below your perfusion pressure threshold. Don't worry though – I've seen dozens of pilots G-LOC without any lasting effects, except for some slight neck pain.

In the early days of manned spaceflight, the direction of acceleration was even more important than it is today because of the sheer magnitude of the acceleration. For example, the Mercury (Figure 3.4), Gemini, and Apollo flights had launch accelerations of 4.5–6.5 +Gx for six minutes and anywhere from 6 to 11 +Gx during re-entry. In fact, NASA was so concerned about the possible effects acceleration might have on the astronauts that they forced them to spend countless hours in the centrifuge (up to 45 hours in some cases; Figure 3.5). Then the Shuttle came along, and astronauts were given a break from the punishing Gs; the now-retired Shuttle had a maximum of *only* 3.2 +Gx during launch and 1.2 +Gz (briefly 2.0 +Gz during turns) during re-entry. Fortunately for the new breed of commercial astronauts, the acceleration forces imposed by most of the current crop of space vehicles should be reasonably comfortable for most, although there will be some who will do better than others.

How you perform in a centrifuge is partly down to the physiological luck of the draw; your tolerance depends on factors such as your tolerance to +Gz acceleration, which is in turn dependent on your height and weight, smoking history, fitness level, hydration, type of acceleration profile, previous and recent exposure to +Gz forces. For example, tall thin people typically don't fare well in a centrifuge because blood has to travel a longer distance to the brain and the eyes. Smokers tend to do well because their arterial beds are less flexible which means it's easier for blood to travel through them. So, if you're a short, squat, chain-smoker, chances are you're ideal centrifuge material! The type of G exposure is important too, because the maximum +Gz level, exposure duration, and the rate of +Gz

3.4 Mercury-Redstone 3 launch. Courtesy: NASA

onset determine the risk of injury to your heart and musculoskeletal system. The most problematic acceleration is rapid-onset rate (ROR), defined as increases greater than 0.33 G per second. ROR tolerance limits are approximately 1 +Gz lower than gradual-onset rate (GOR) tolerances because they exceed the ability of the cardiovascular system to get enough blood to your brain. RORs can also result in the dreaded G-LOC without any

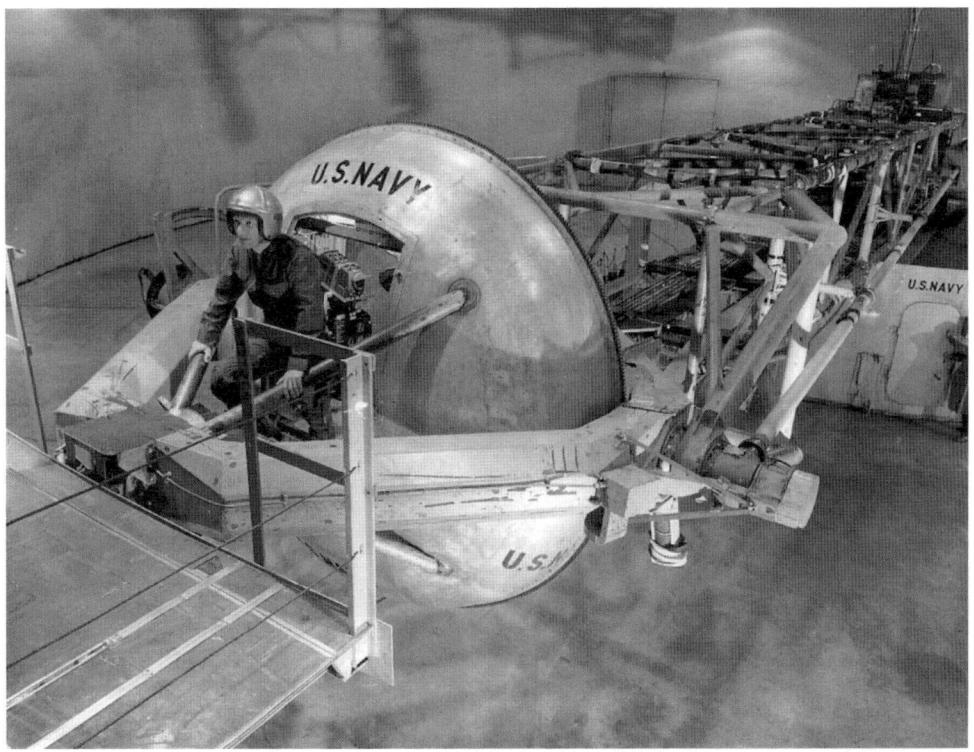

3.5   The Johnsville centrifuge. Courtesy: NASA/US Navy

visual warning symptoms such as tunnel vision, gray-out, or black-out. To prevent this happening when they're performing aerobatic maneuvers, fighter pilots wear anti-G-suits (Figure 3.6) which increase their G-tolerance by up to 1.5 +Gz. Another way fighter pilots increase their G-tolerance is by using the anti-G straining maneuver (AGSM), which can increase tolerance by as much as 3 +Gz.

Over the years, centrifuge data have allowed scientists to develop a model of +Gz tolerance limits which incorporate +Gz magnitude, duration, and rate of onset (generally, with no protection, most healthy people can tolerate up to 4 +Gz acceleration for ROR profiles and up to 4.5 +Gz with GOR profiles). Okay, so that's +Gz tolerance, but what about −Gz and the transition from one type of G to another? This transition is where most problems occur because transition to +Gz can cause a big drop in your brain's blood pressure. That's bad news for the cardiovascular system because it can take quite a while before your body reacts and compensates. In fact, if you're exposed to −Gz before transitioning to +Gz, you can become a victim to the "push–pull effect". This can be deadly. Usually, this "push–pull effect" occurs in combat engagements and has been implicated in several fatalities. Even now, with a whole library of G data, there is still a knowledge gap of this issue and no countermeasures have been developed. It's unclear whether a "push–pull effect" will occur in transition from microgravity to re-entry deceleration, but it has been described in

3.6 The author being fitted for his G-suit in preparation for his Hawk flight. Courtesy: author

parabolic flight and there are some who are concerned it could occur in suborbital flights. That's because, in suborbital flight, the "push–pull effect" is prolonged by increasing the duration of the prior −Gz exposure. Normally, the −Gz exposure is only a few seconds in combat flight whereas in parabolic flight the exposure is 20–30 seconds. But what about after four minutes of suborbital flight? Truth is, we just don't know whether four minutes of microgravity will provoke the same response or cause a further deterioration in +Gz tolerance.

If you happen to be one of the lucky ones with a suborbital flight ticket and you're worried about how you might be affected by G, you should know that the acceleration envelope recommended by the Federal Aviation Administration (FAA) for commercial aerospace vehicles should not exceed +3 Gz (−2 Gz), ±6 Gx, and ±1 Gy. If you experience these levels as GORs, you shouldn't have a problem as long as you're healthy. If you happen to be flying in the 2014 to 2016 timeframe chances are you'll be flying on SS2 in which case you have an idea of what to expect. During SS2's rocket engine boost, you will experience acceleration as high as +3.8 Gx followed by a brief spike up to +4.0 Gz as the vehicle rotates to a nose-high attitude. On re-entry, the pilot's will feel 6 G mainly in the +Gz axis but, thanks to SS2's tilt-back seating, most of the acceleration you will feel will be in the Gx axis. Duration of these G-forces is expected to be about 70 seconds during launch and about 30 seconds during re-entry.

Microgravity Effects

Like acceleration, the physiological changes resulting from exposure to microgravity vary from individual to individual. While suborbital microgravity exposure will only last for four or five minutes, if you happen to be inexperienced, non-adapted, or highly sensitive, chances are you will experience neurovestibular and/or cardiovascular symptoms. Although no proof exists, parabolic flight experience might be a way to alleviate suborbital flight symptoms by providing weightless experience, which is why Suborbital Training recommends zero-G training as part of astronaut training.

Cardiovascular Effects

A common cardiovascular effect observed in Shuttle astronauts while they were lying down awaiting launch was a shift in fluids from their legs to their head. Part of the reason for this was due to the slightly head-down pre-launch position. In orbit, due to the absence of gravity, fluids shifted again, with body fluids rushing to the head, giving astronauts a sensation of head fullness. Fortunately, most of these effects won't be a problem during your suborbital flight simply because these physiological changes take time to develop in microgravity.

Neurovestibular Effects

In common with fluid shifts, neurovestibular effects won't be much of problem for most suborbital passengers because these effects also take time to manifest themselves. After orbital flight, astronauts suffer an altered ability to sense tilt and roll, defects in postural stability, impaired gaze control, and changes in sensory integration. Basically, they are discombobulated! While suborbital passengers probably won't be affected to the degree that astronauts returning from orbit are, there have been neurovestibular alterations observed in even short exposures to zero-G. For example, illusions and disorientation were reported on several X-15 flights. Now, you may think why not test this in a simulator, but the problem is that rapid launch acceleration followed by zero-G followed by re-entry deceleration can't be tested in continuity. Another problem is that many commercial astronauts will be flying these suborbital profiles repeatedly and (the really lucky ones) maybe even on a daily basis and there isn't any experience that shows whether repeated and frequent suborbital profile exposures will be adaptive or maladaptive to neurovestibular function. Obviously, more experience in suborbital flight is needed and it will be the job of the new corps of astronauts to determine whether this is even an issue.

X-15 Neurovestibular Experience

Although we don't have much suborbital flight experience, it's worth taking a look at the experience of those pilots who flew similar flights back in the 1950s and 1960s. The most notable suborbital flights were flown by the X-15, three of which were built for a program that comprised 199 flights; the first was flown on June 8th, 1959, and the last on October 24th, 1968.

Among the 12 pilots of the X-15 program were future NASA astronauts Neil Armstrong and Joe Engle. During the program, eight pilots met the USAF spaceflight criteria by exceeding an altitude of 80 kilometers, thus qualifying them for astronaut status. Of all the X-15 missions, only two flights[4] (piloted by Joe Walker) qualified as spaceflights according to the FAI definition. Although biomedical data weren't measured on the X-15 flights, pre-flight and post-flight flight surgeon examinations were performed and nothing unusual was reported. Also, pilot performance wasn't impaired by launch acceleration followed by zero-G followed by re-entry (don't forget these were phenomenally experienced pilots – the very best on the planet), although the G-forces imposed on the pilot during the boost phase of the flight often resulted in severe vertigo. In fact, vertigo was a serious problem throughout the program and was blamed for the death of Major Mike Adams. Major Adams's seventh flight took place on November 15th, 1967. At 10:30 in the morning, his X-15 dropped away from the NB-52B at 13,700 meters and, at 10:33, he reached a peak altitude of 81,000 meters. On reaching its maximum altitude, the X-15's heading was off by 15° and, as Adams descended, the aircraft began drifting to the right. Then, at 70,000 meters, encountering rapidly increasing dynamic pressures, the X-15 entered a Mach 5 spin.

NASA controllers advised Adams he was "a little bit high", but in "real good shape". Adams replied that the aircraft "seems squirrelly". At 10:34 came a troubling call: "I'm in a spin." With no heading information, controllers saw only large and very slow pitching and rolling motions and thought Adams was overstating the case. But Adams radioed again: "I'm in a spin." Unfortunately, there was no spin recovery technique for the X-15, and engineers knew next to nothing about the aircraft's supersonic spin behavior. The chase pilots, realizing Adams would never make Rogers Dry Lake, raced for the emergency lakes. Somehow, the aircraft recovered from the spin at 36,000 meters and went into an inverted Mach 4.7 dive. Adams now had a good chance of rolling upright, pulling out, and landing. But then the aircraft began a rapid pitching motion diving at 49,000 meters per minute. Dynamic pressure increased intolerably and, as the aircraft neared 20,000 meters, it was diving at Mach 3.93 and experiencing over 15 G vertically, both positive and negative, and 8 G laterally. The aircraft broke up 10 minutes and 35 seconds after launch.

---

[4]   Flight 90, on July 19th, 1963, reached 105.9 kilometers. Flight 91, on August 22nd, 1963, reached 107.8 kilometers.

NASA and the Air Force convened an accident board, which concluded Adams[5] had suffered severe vertigo during climb-out which caused spatial disorientation. Small heading deviations caused by a degraded flight control system were made worse by incorrect pilot inputs at an altitude of over 20 kilometers. The board concluded Adams had misinterpreted a roll indication for a slide slip indication and had made control inputs in the wrong direction. As a result of the accident investigation, it was recommended all future X-15 pilots be medically screened for labyrinth (vertigo) sensitivity.

Space Motion Sickness

Do you get motion sick? Well, not to worry, because there is little or no correlation between being sick on Earth and being sick in space. In fact, some people who are chronically motion sick on Earth are just fine during a parabolic flight. Equally, those who have never experienced terrestrial motion sickness are sometimes sick as the proverbial dog when they reach orbit. No doubt about it, space motion sickness (SMS) is a problem. In fact, more than 70% of first-time astronauts flying orbital spaceflights suffer from it; the syndrome, thought to be due to a sensory conflict between visual, vestibular, and proprioceptive stimuli, has been a problem as long as there have been astronauts. Symptoms typically occur within the first 24 hours, but some astronauts have reported symptoms – dizziness, pallor, sweating, severe nausea, and vomiting are the most common – immediately after main engine cut-off. Vomiting, which can be especially messy in zero-G, can crescendo suddenly without any warning symptoms. And, in a multi-passenger vehicle, one passenger becoming nauseated can potentially trigger nausea in the other vehicle occupants – just imagine trying to do your work as a commercial astronaut or trying to take pictures as a space tourist while barfing into a vomit bag or trying to dodge balls of stomach contents as they are ejected from your fellow passengers! It's an image the commercial suborbital operators have played down, and for good reason because, after spending US$250,000 on a ticket, nobody wants to have their once-in-a-lifetime experience spoilt by projectile vomiting. Of course, anti-motion sickness medications can be used, but these tend to have side effects which aren't always conducive to enjoying spaceflight.

Unfortunately, there is just no way that will guarantee you won't be sick. Parabolic flight adaptation and experience in high-performance jet aircraft don't work. Neither do rotating chairs nor centrifuge training. But there is some success with the use of training aids that duplicate the sensory conflict that occurs in parabolic flight. And, if you want to take advantage of this training, you best bet is to visit **SIRIUS** Astronaut Training (www.siriusastronauttraining.com) in Boston. Run by human spatial orientation specialists James Lackner, Paul DiZio, and Janna Kaplan,[6] SIRIUS focuses on the sensorimotor human factors of spaceflight such as motion sickness, spatial disorientation, spatial illusions, and

---

[5]   Mike Adams was posthumously awarded Astronaut Wings for his last flight in the X-15-3, which attained an altitude of 81,070 meters. In 1991, his name was added to the Astronaut Memorial at the Kennedy Space Center in Florida.

[6]   Janna can be reached at janna.kaplan@siriusastronauttraining.com (phone: 781-736-2038).

movement errors in changing gravitoinertial force environments. Chances are, if you take their course, you'll stand the best chance of having a vomit-free flight.

Emergency Egress Capability

One of the most important medical and training assumptions is that passengers will be capable of performing an emergency egress without assistance. Obviously, performing an emergency egress will depend on the type of vehicle you're flying because some vehicles don't have such a capability, but that's another story. Back in the days when the US government had its own manned spaceflight capability, up to 15% of Shuttle astronauts were judged too impaired post landing to perform an unaided egress. The reason Shuttle astronauts were so discombobulated was due to a combination of re-entry motion sickness and post-flight orthostatic intolerance. Fortunately, as some of these problems are dependent on the duration time of microgravity, they shouldn't be a problem after suborbital flights, but there may be some susceptible individuals who will have issues. It's one of the reasons why companies such as Suborbital Training recommend emergency egress training (Figure 3.7).

3.7 Dunker training. Courtesy: A4H

Environmental Medical Issues

You may be wondering how cabin environment can affect your performance, but a malfunction of the cabin heating, air circulation, and/or cooling systems will seriously affect how you think and how you manipulate objects, such as that expensive Nikon Digital SLR. Also, SMS is known to be exacerbated by over-heating, and then there's the issue of cabin pressure. In common with cabin temperature and humidity (which should be around 21–26°C with a relative humidity of 30–40%), cabin pressure will vary depending upon the design of the space vehicle. In current orbital space vehicles, the cabin pressure is maintained at a sea-level pressure of 14.7 psi (101 kPa), allowing for a shirt-sleeved environment.

Any vehicle that ventures beyond the Armstrong Line[7] will operate at such high altitudes that there is also a risk of a rapid or explosive decompression, either of which could result in hypoxia and/or death due to *hypoxia* or *ebullism*. In case you're wondering how bad such an event might be, here's what *The USAF Flight Surgeon's Guide* has to say about some of the effects due to mechanical expansion of gases during rapid decompression:

Gastrointestinal Tract during Rapid Decompression

One of the potential dangers during a rapid decompression is the expansion of gases within body cavities. The abdominal distress during rapid decompression is usually no more severe than that which might occur during slower decompression. Nevertheless, abdominal distension, when it does occur, may have several important effects. The diaphragm is displaced upward by the expansion of trapped gas in the stomach, which can retard respiratory movements. Distension of these abdominal organs may also stimulate the abdominal branches of the vagus nerve, resulting in cardiovascular depression, and if severe enough, cause a reduction in blood pressure, unconsciousness, and shock.

Sounds painful doesn't it? Well, rapid decompression is probably more survivable than explosive decompression – an event that can cause your blood to literally boil. Here's what Dr. Tamarack R. Czarnik, a specialist in aerospace medicine, has to say about the medical effects of an explosive decompression:

Damage to the lungs in rapid or explosive decompression occurs primarily due to pulmonary overpressure, the tremendous pressure differential inside versus outside the lungs. 80 mm Hg is enough to cause pulmonary tears and alveolar rupture; pulmonary hemorrhaging, ranging from petechiae to free blood is also seen. Emphysematous changes are seen especially in the upper lungs, while atelectasis

---

[7]   The Armstrong Line (18,900–19,350 meters) has nothing to do with Neil Armstrong. The "line" is named after Harry George Armstrong, who founded the US Air Force's Department of Space Medicine, and it represents the altitude that produces an atmospheric pressure so low that water boils at normal body temperature. In fact, if you were exposed to an altitude above the line, your exposed bodily liquids would simply boil away and you'd be dead within a minute or two.

3.8 The author prepares for a high-altitude indoctrination chamber flight to 7,620 meters.
Courtesy: Chris Townson

and edema predominate in the lower lungs. When we get to the patient, the lungs
will be a bloody, ruptured mess.

Though these are the most life-threatening changes seen in ebullism, subcutane-
ous swelling is also seen, due to creation of water vapor under the skin. This can
rapidly distend the body to twice its normal volume. Our patient will look no better
than he feels, though this means little in terms of survival.

Bleak, isn't it? The best way to avoid these nasty hypobaric events is to use a pressure
suit. Until very recently, pressure suits were expensive and tended to weigh a lot: the
Shuttle pressure suit weighed over 20 kilograms and cost more than US$100,000.
Fortunately, for the new crop of suborbital passengers, pressure suits will be a lot less
bulky and, for the operators, less expensive. For example, Final Frontier Design (FFD) is
planning on selling its pressure suit for around US$20,000. Nevertheless, even if you are
wearing a pressure suit, it's useful to understand the consequences of one failing, which is
why operators include high-altitude indoctrination (Figure 3.8) to familiarize passengers
with the hazards of hypoxia and depressurization.

Radiation

Radiation in space – ionizing radiation – consists of subatomic particles that can interact
with biological tissues and destroy DNA strands, causing genetic damage that can in turn
lead to dangerous mutations. Sources of this type of radiation in space include galactic
cosmic radiation (GCR), solar radiation, solar flares, and trapped radiation from the Van

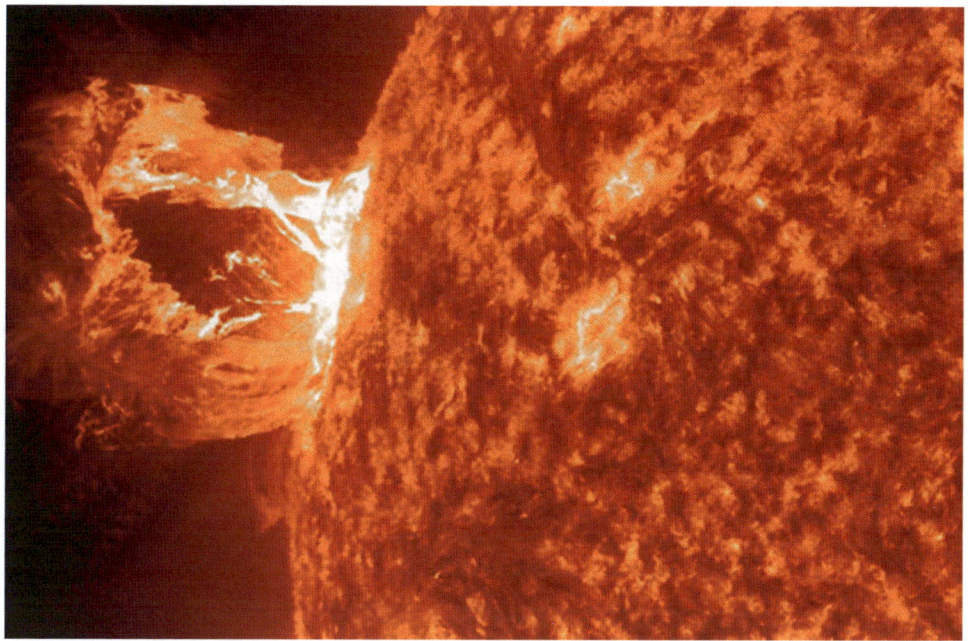

3.9   A solar flare. Courtesy: NASA

Allen belts. GCR originates outside of the Solar System and consists of hydrogen nuclei protons (87%), helium nuclei alpha particles (12%), and damaging high-energy heavy nuclei such as iron (1%), while solar cosmic radiation (SCR) comprises proton–electron plasma ejected from the surface of the Sun. Completing the radiation cocktail are solar flares (Figure 3.9), which are magnetic disturbances on the Sun's surface generating electromagnetic radiation.

How much radiation is too much? Well, the dose standard for radiation-exposed workers is 20 mSieverts (Sv) per year (averaged over five years) and an exposure to this level over 40 years results in an excess lifetime fatal cancer risk of 3.2%. By comparison, orbital spaceflight results in a variable radiation dose exposure depending on the orbital altitude and solar activity. On average, astronauts on board the International Space Station (ISS) are exposed to about 0.01–0.1 Sv a month. Fortunately, radiation levels at suborbital flight altitudes won't be anything close to those levels. In fact, when you fly on your suborbital flight, you won't be exposed to a radiation level much higher than a Concorde flight, which will equate to less than 15 microSv per hour. It doesn't sound a lot but if you happen to be a commercial suborbital astronaut, chances are you'll be flying multiple missions, so what are the recommended occupational exposure limits for this new employee? Well, the International Commission on Radiological Protection (ICRP) says that for commercial aircrews, the occupational limit is 20 mSv per year, averaged over five years, with a maximum in any one year of 50 mSv. Compare this to the ICRP recommendation for the general public that states exposure should be less than 1 mSv per year. Another advantage commercial suborbital astronauts will have over their government employed co-workers will be that suborbital flight schedules can easily be altered in the event of unfavorable atmospheric conditions such as a solar storm.

Noise

Launching a rocket – any rocket – is a noisy business, and vehicles being launched into suborbital space require powerful thrust that happens to be loud. Very loud. And this noise is transmitted through the whole vehicle and, because the vehicle is an enclosed space, this noise is reflected multiple times off the walls, bulkheads, floors, and ceilings. Although the noise levels are relatively short, the magnitude can be quite intense – so intense that passengers may suffer reduced visual acuity, vertigo, nausea, disorientation, and ear pain. Noise levels in the crew compartment during a Shuttle launch reached almost 120 dB (equivalent to the sound of a Iron Maiden concert in front of the speakers). Because of this assault on your hearing, auditory protection will definitely be required during a suborbital launch.

Vibration

As well as all that noise, the power being unleashed to launch your vehicle will generate an awful lot of vibration (watch in-cabin videos of the SS1 flights during ascent and you'll see what I mean). How much? Think about the vibration you feel when an aircraft takes off and multiply that by several orders of magnitude and you'll have some idea. While vibration won't be more than a mild and temporary inconvenience for fare-paying passengers, for commercial astronauts tasked with flying payloads, it could be a problem. That's because vibration can cause manual tracking errors and can interfere with ability to visually track displays, which could be a problem for an astronaut tasked with keeping an eye on an experiment.

Suborbital Medical Standards

So what do these risks mean in terms of how healthy you need to be to be medically qualified for suborbital flight? Well, the FAA[8] thought about this and appointed a team to evaluate the medical standards that would be appropriate for suborbital passengers. The report, led by principal investigator Dr. Richard Jennings, and co-investigators Drs. James Vanderploeg, Melchor Antunano, and Jeffrey Davis, is titled *Flight Crew Medical*

---

[8]   According to the FAA, a suborbital operator must: (1) ensure that any crew-training device used to meet the training requirements realistically represents the vehicle's configuration and mission or (2) inform the crew member being trained of the differences between the two and (3) maintain training records. An operator must continually update the crew training to ensure it incorporates lessons learned from training and operational missions. An operator must: (1) track each revision and update in writing and (2) document the completed training for each crew member and maintain the documentation for each active crew member and (3) establish a recurrent training schedule and ensure all crew qualifications and training required by § 460.5 are current before launch and re-entry.

Standards and Spaceflight Participant Medical Acceptance Guidelines for Commercial Space Flight, and it was published on June 30th, 2012. What follows is a synopsis of the report.

The FAA-sponsored medical guidelines project was conducted in three phases, the first of which collected and reviewed documents addressing suborbital crewmember and space-flight participant (SFP) medical certification. In the second phase, a preliminary document incorporating the guidelines and recommendations as outlined in Phase I was prepared and a working group of experts was convened to consider the comments from Phase II. Then, in Phase III, a consolidated set of recommendations for the medical certification of crewmembers, medical acceptance guidelines for SFPs, and recommended training procedures was prepared and the document was provided to the FAA.

The first part of this document outlined a reference mission, which set out a number of assumptions. The first of these was that a suborbital spacecraft will provide a shirt-sleeve cabin environment with an appropriate temperature, a cabin pressure of not more than 2,400 meters equivalent altitude, and appropriate oxygen and humidity levels. The second assumption was that the acceleration in a suborbital spacecraft should not exceed +6 Gx, +1 Gy, and +4 Gz. If the acceleration profile exposed SFPs to greater than +4 Gz, then the SFPs should be medically screened according to the guidelines outlined for orbital passengers. The third assumption dealt with flight rates and assumed SFPs will only participate in one suborbital flight per day, whereas commercial astronauts or flight crew could make multiple flights per day. The document also noted that repeated flights to the acceleration limits listed, with four minutes of zero-G exposure between launch and entry, haven't been performed before, so caution should be exercised until an experience base is acquired. Finally, the document assumed that the radiation dose will not exceed the yearly commercial airline passenger dose, defined as no more than 1 mSv per year.

The next part of the document dealt with the guidelines for screening. The guidelines suggested that the content and extent of a medical questionnaire and physical exam should be related to each operator's flight profile and that SFPs should complete a medical questionnaire (see Table 3.1) and a physical exam by a qualified physician with knowledge of the spaceflight environment.

In addition to completing the questionnaire, you will be required to inform the doctor if you have a medical condition that would impair your ability to safely perform a suborbital flight without compromising the safety of other occupants and/or safely perform an emergency egress without assistance. And, given the novel flight environment, a post-flight medical debrief is recommended to collect post-flight medical data, and enquire about health effects of the flight. After completing the questionnaire, prospective suborbital astronauts will be medically screened. According to the project's findings, the medical screen may determine that a potential SFP has a medical problem requiring additional consideration. While there are no hard and fast suborbital medical standards, there are some conditions that could be cause for concern. For example, any condition that may result in an in-flight death or injury is obviously a red flag. Also, a person that has a condition with functional defects that could interfere with the use of personal protective equipment or interfere with an emergency egress probably shouldn't be sold a ticket. Another medical issue is any problem that may be exacerbated by the operational environment or flight-related stress. So what do you do if you don't meet the recommended

Table 3.1. Spaceflight Participant Questionnaire.
If you're planning on taking a suborbital flight, chances are you will need to indicate a history of any of the following conditions:

Otitis, sinusitis, bronchitis, asthma, or other respiratory disorders	Mental disorders, anxiety, or history of hyperventilation
Dizziness or vertigo	Claustrophobia
Fainting spells or other loss of consciousness	Attempted suicide
Seizures	Use of medications
Tuberculosis	Alcohol or drug dependence or abuse
Surgery and/or other hospital admissions	Current pregnancy or recent spontaneous or voluntary termination of pregnancy
Visits to health care provider in last 3 years	Recent significant trauma
History of decompression sickness (DCS)	Diabetes
Anemia or other blood disorders	Cancer
Heart or circulatory disorders, including implanted pacemaker or defibrillator	Rejection for life or health insurance
Disability or deformity requiring accommodation	

guidance criteria? One option is a mitigation strategy, although the operator and aerospace medicine physician must ensure that the condition and the mitigation process won't impair the ability of the SFP to safely perform activities required for the flight including an emergency egress. Part of this mitigation strategy may involve training, which forms the subject of the second part of this chapter.

TRAINING FOR COMMERCIAL SUBORBITAL SPACEFLIGHT

> "I am living in fear of the move to develop standards for crew and passenger training in this industry. I think it's a mistake at this stage in the development of the industry."
> Jeff Greason, CEO of XCOR Aerospace

Training or high-performance environment exposures for SFPs may be part of a medical risk mitigation strategy[9] but it is also necessary for everyone planning on becoming a suborbital astronaut. Those of you who have watched videos of Virgin Galactic passengers completing their training will be familiar with centrifuge training and high-altitude

---

[9] SFPs with medical deficiencies may be medically monitored during exposure to analog environments as part of the risk mitigation strategy. Also, SFPs may elect to participate in medical flights where more extensive medical capabilities are available. In the event a risk remains, the physician and operator will provide the best information available at the time for informing the SFP of that risk. It is possible that medical conditions determined to be unstable (such as uncontrolled hypertension or angina) can be successfully treated to permit a SFP to be approved for a flight.

indoctrination, but there is a lot more preparation specific to the suborbital spaceflight environment.

That's not to say suborbital passengers should undergo the comprehensive training government astronauts complete. After all, astronauts employed by government space agencies are trained for orbital flight, which is much more rigorous than suborbital flight. Also, orbital missions nowadays tend to last several months whereas a typical suborbital mission lasts no longer than a couple of hours depending on the mission architecture. So, for suborbital spaceflight, there's no need for much training, but how much is appropriate? It's a question that has yet to be answered because the industry has not yet standardized astronaut training. Even the FAA has provided only vague guidelines as to what suborbital flight training should include. Fortunately, there are a number of training options available for SFPs and those hoping to be employed as commercial astronauts. For example, the NASTAR Center was used by Virgin Galactic to train some of its SFPs, so it's worth taking a closer look at this facility.

NASTAR

The NASTAR Center is a non-government, world-class aerospace training facility that supports the training, research, and educational needs of the aerospace industry. Established in 2006, the center began as a product showcase, engineering development and test center for its owners, the Environmental Tectonics Corporation (ETC), but soon became recognized for its unique approach and sophisticated interactive flight training technology. The center contains all sorts of space training equipment, ranging from high-fidelity simulators to a multi-axis centrifuge (Figure 3.10) – the ATFS-400 Phoenix.

In June, 2009, shortly after Virgin Galactic started using NASTAR for their SFP training, Alan Stern and Dan Durda of the Southwest Research Institute (SwRI) paid a visit with the aim of establishing a training program for suborbital scientists. In August 2009, SwRI and NASTAR established the Suborbital Scientist Training (SST) course, a two-day program designed to qualify individuals for the physiological rigors of suborbital spaceflight and to prepare them for suborbital Research and Education Missions (REM):

> "NASTAR is the leader in private space flight training. Having trained for and flown high-performance F-18s and high-altitude reconnaissance jets like the WB-57, I could see that NASTAR was uniquely positioned to support this program with its extensive, modern facilities and their application to suborbital research in space by scientists, engineers, educators and even students."
>
> Alan Stern, associate vice president of the SwRI Space Science and Engineering Division, commenting on NASTAR's Suborbital Scientist Training program

On Day #1 of the SST program, students are given a tour of the facility during which they are shown the ejection seat simulators, the yaw, pitch and roll confusion generators, hypobaric chamber, and, of course, the centrifuge, complete with spaceflight simulator pod. Then, the budding astronauts are fitted for flight suits and instructed on the basics of aerospace physiology. In the afternoon, they complete their high-altitude indoctrination inside the hypobaric chamber while attempting to solve problems on a worksheet to test their reaction to hypoxia.

3.10   The NASTAR centrifuge. Courtesy: NASTAR

Day #2 begins with prepping for the G-tolerance test, which includes teaching the students the AGSM, which helps combat G-forces. After practicing their AGSM technique, students perform four centrifuge runs ("Flights" in NASTAR parlance). After enjoying/surviving the centrifuge, students participate in a distraction exercise which takes place in a room with a black interior and set up with a Virgin Galactic SS2-sized cabin area. During the exercise, a projector plays space images as if viewed out of one of SS2's windows. At the front of the simulated cabin are six boxes representing experiments. The task for the students is to open their assigned boxes and perform the experiment within five minutes, which happens to be the period of zero gravity time. The experiments usually include tasks such as matching patches around the cabin with patches in the box to finding odd items in a picture. Students typically find the whole exercise a discombobulating experience the first time around, although by the second attempt they get smart and develop a strategy. After matching patches, the students experience a simulated launch and re-entry sequence in SS2.

A few of my space industry colleagues have taken NASTAR's SST course and they say it's a blast. But, does two days of training really qualify you to be a SFP or a scientist astronaut? After all, there is no emergency egress training in NASTAR's program, so what happens if you fly on board a vehicle that suffers a major malfunction and you need to bail out? And, after bailing out, what happens if you find yourself in the ocean? Might not survival training be useful? Well, there are a few organizations thinking along the lines of offering more comprehensive training programs. One of these is A4H.

Astronauts for Hire

A4H (Figure 3.11) is one of the few organizations that have taken the first step to standardize commercial astronaut training by developing a qualification standard for its members. Now, A4H didn't set standards for the sake of setting standards, nor, as some people (misguided journalists for the most part) think, is it trying to impose those standards on the industry. All it's trying to do is provide a safeguard for its members and value for money to its clients. After all, research institutions and other potential employers will require that their employees flying their payload possess the knowledge and skills necessary to perform the many tasks required of a scientist in the suborbital research environment. While vehicle-specific training will no doubt be delivered by suborbital launch providers, who will be responsible for providing the training for the scientists? Well, in the case of A4H, the scientists will be trained in accordance with A4H's in-house suborbital scientist training program.

A4H decided each of their flight members should be trained to operate safely within the vehicle so they will know how to respond to planned *and* anomalous events. Prior to each mission, A4H flight members will receive vehicle and mission-specific training to cover all phases of flight by practicing the operational and procedural simulations of each mission phase. Flight members are also required to be familiar with the flight characteristics

3.11 Members of Astronauts for Hire. From left to right: José Hurtado, Brian Shiro, Alli Taylor, Kristine Ferrone, Jason Reimuller, Jules, Chris Altman. Courtesy: A4H

3.12   Unusual attitude flying as performed by a pair of NASA T-38s. Courtesy: NASA

of the vehicle so they are proficient in nominal and non-nominal flight conditions (such as abort scenarios and emergency operations). They also must have a competent understanding of generic vehicle systems, vehicle characteristics, and vehicle capabilities, as well as operational, malfunction, and contingency procedures.

In addition to the NASTAR course, A4H flight members must complete a zero-G flight, and the PADI or NAUI Open Water Dive Course. To demonstrate their proficiency in emergency egress, flight members are required to complete basic egress training such as the helo dunker, which teaches them how to react in a timely manner while under stress (being drowned!). Another stressful training increment is aerobatic or unusual attitude training (Figure 3.12) which teaches flight members how to adapt to rapidly changing

G-loads. In between completing their training, A4H members work towards completing four core academic modules: Human Factors in Space, Life Support Systems, Spacecraft Systems Engineering, and the Space Environment. Once they've checked all the boxes, they're flight-eligible.

Suborbital Training

Another company in the business of offering suborbital training and providing customized space training is Suborbital Training (www.suborbitaltraining.com; contact e-mail: suborbitaltraining@hotmail.com). As a provider of suborbital training services, the company provides affordable, discounted (thanks to agreements with several companies across the industry), and mission-specific suborbital flight training customized to the client's needs, whether you happen to be a SFP or a scientist. Among its many services (see Table 3.2), the company specializes in the delivery of suborbital flight academic training modules and provides pre-mission, in-mission, and post-mission biomedical support. For scientists, this support crosses the spectrum of the training required to be an effective and proficient suborbital payload specialist, specifically Vehicle Familiarity, Environment Training, Payload and Science Familiarity, and Interpersonal Training. It's a small but highly versatile commercial spaceflight company that has an extensive background in the space life sciences arena and fully understands the challenges of the theoretical and practical training required for suborbital flight. By customizing the client's training requirements, Suborbital Training aims to provide its customers with a high-quality, time-saving product that is designed to ensure mission success.

Inner Space Training

Inner Space Training is based in the Netherlands (www.innerspacetraining.com; contact e-mail: mindyhoward@innerspacetraining.com) and caters to a specific niche of training: ensuring suborbital astronauts get the most of their spaceflight experience. How? It's all down to training passengers to relax and be calm during the mission. The one-day course prepares clients for the psychological and emotional challenges of different flight phases, focuses their mental acuity and concentration to help them accomplish their mission objectives, and helps passengers create strategies to deal with relationships with other passengers to resolve any potential conflicts. During the training (Table 3.3), participants undergo various personal and team exercises to build their understanding of what the suborbital space environment entails and how their mission objectives can be achieved. Along the way, participants learn what the various flight phases are and become aware of any physical and psychological challenges, along with finding solutions to them. After the course, participants practice their spaceflight, training their mind using Brainwave Entrainment (BWE) technology. The techniques and the positive effects of the BWE program can be reactivated during spaceflight to give astronauts calm focus during their entire flight, enabling them to have a peak experience and accomplish their missions.

For those interested in this sort of training, you can book through Suborbital Training with whom Inner Space Training has a 20% discount.

Table 3.2. Suborbital Training.

Service	Description
Vehicle Familiarity Training Program	This program features academic instruction on emergency procedures, instruction on intravehicular orientation, and an overview of basic spacecraft systems and sub-systems. The program comprises the following sub-modules which can be taken individually or collectively: (i) Instruction on the anti-G straining maneuver, (ii) High-altitude indoctrination at a facility nearest to your location, (iii) Emergency Egress Training (vi) Generic zero-G exercises.
Environment Training Program	This program provides clients with the theoretical and practical aspects of training to become a suborbital payload specialist. The program comprises the following sub-modules which can be taken individually or collectively: (i) G-Physiology (ii) High-Altitude Physiology, (iii) Emergency Egress (iv) Preflight Preparation (v) In-flight indoctrination (vi) Survival Training Theory (vii) Flight Vehicle Systems Theory (viii) Post-flight preparation.
Payload and Science Familiarity Program	This program provides scientists with the theoretical and practical aspects of flying your payload and the steps necessary to successively fly their science experiments, including customized cue cards. This program comprises the following sub-modules which can be taken individually or collectively: (i) Payload integration, (ii) Pre, in, and post-flight science protocols, (iii) Payload checkout and validation procedures.
Interpersonal Training Program	This program provides scientist astronauts with instruction on the latest in space crew resource management techniques. This program comprises the following sub-modules which can be taken individually or collectively: (i) Information Processing for Suborbital Payload Specialists, (ii) Human Error and Error Management, (iii) Situational Awareness, (iv) Communication and Management.
Suborbital Training Preflight Fitness Program	This program is based on research-validated exercises proven to increase a person's tolerance to Gs. The program was devised by a trained commercial suborbital astronaut and Director of Canada's Manned Centrifuge Operations.
Corporate Suborbital Payload Specialist	Don't have the time to be trained for a suborbital flight but are planning on flying a payload? No problem; Suborbital Training can fly it for you. Suborbital Training has a fully trained corporate suborbital astronaut on staff with extensive scientific research experience who can take care of your payload and/or science experiment. Their astronaut served as the Director of Canada's Manned Centrifuge Operations between 2009 and 2012 and served as Director of Hypobaric Operations at two hypobaric facilities. He has performed dozens of centrifuge rides, scores of high-altitude indoctrination runs to 25,000 feet, and has performed unusual attitude training in the BAE Hawk aircraft. Furthermore, Suborbital Training's commercially qualified astronaut is a qualified free-fall parachutist, who is trained in emergency egress, sea, desert and jungle survival, and also happens to be a qualified scuba diver with more than 250 dives to his credit. In addition to being a veteran of ESA's Parabolic Flight Campaign, Suborbital Training's astronaut in residence is a qualified private pilot, has a Master's degree, Ph.D., and a post-doctoral degree.

Table 3.3. Topics Included in Inner Space Training.

Psychological acclimation and anchoring
Peak experience and conditions to create one
Intention setting for your journey
Situational Awareness Flow Protocol (theory)
Situational Awareness Flow Protocol (practice)
Understanding Spacecraft Culture
Becoming a Social Ambassador
Brain Wave Entrainment (BWE) – theory and practice for optimizing your "inner space" using the MindSpa® device

SIRIUS Astronaut Training

While Inner Space Training focuses on the psychological aspects of spaceflight, SIRIUS Astronaut Training (www.siriusastronauttraining.com; contact e-mail: janna.kaplan@siri-usastronauttraining.com) develops programs that help passengers to deal with the problems of SMS and spatial disorientation. Based on their long-standing involvement in space research and employing unique research facilities at the Ashton Graybiel Spatial Orientation Laboratory (AGSOL), SIRIUS has developed programs that help passengers adapt to spaceflight through the use of gravity/rotating environments and motion platforms. Take their Multi-Axis Rotation and Tilt (MART) device, for example. This device allows SIRIUS to expose passengers to motion in two axes simultaneously, the result of which is spatial disorientation, which usually means clients have trouble determining "up" and "down". But, thanks to being exposed to this disorientation, and thanks to SIRIUS's carefully designed protocols, spending time in the MART device helps passengers pre-adapt. Another useful piece of equipment used by SIRIUS is the Vection Chamber. The black-and-white striped walls and floor of this cylindrical chamber can rotate independently of each other, causing illusory experiences of self-motion in about 60% of clients. SIRIUS uses the device to train passengers to recognize such illusions and be aware that they may impair their actions in spaceflight.

4

The Frontrunners

VIRGIN GALACTIC

On April 29th, 2013, SpaceShipTwo (SS2), a spaceship financed by Sir Richard Branson, made its first powered flight over Mojave, California. Although SS2 didn't actually fly in space during the test flight, it marked a significant milestone for the company that will kick-start the suborbital industry. During the early morning flight, SS2, strapped beneath its mother ship, took off from a runway in the Mojave Desert north of Los Angeles. Once it had reached the release altitude, the carrier vehicle released SS2, which ignited its engine for 16 seconds, before gliding to a safe landing. Although only 16 seconds of the vehicle's 13-minute flight were powered, the test moved Virgin Galactic one significant step closer towards its goal of flying passengers into suborbital space. Coming two and a half years after SS2's first glide flight, the powered test marked the beginning of the envelope expansion phase and the very real possibility of flights into space by the end of 2013, albeit test flights.

SS2 Powered Test Flight

The test began with SS2 mounted underneath the wing of the WhiteKnightTwo (WK2) carrier aircraft, which took off at 7:03 a.m. local time. Launch took place 45 minutes later from an altitude of 14,325 meters. Following release of SS2, test pilots Mark Stucky and Mike Alsbury lit the rocket, which boosted the vehicle to Mach 1.22 and a maximum altitude of 17,130 meters following the 16-second rocket burnout.

Until the April 2013 flight, SS2 had only performed unpowered glide flights, but the first powered flight was without any doubt the company's single most important test to date. Virgin Galactic Founder, Sir Richard Branson, was in Mojave to witness the occasion and was happy with what he saw, predicting passenger flights would soon follow. Sir Richard had predicted commercial flights would begin in 2007, but a deadly explosion during ground testing and longer-than-expected test flights pushed that deadline back.

E. Seedhouse, *Suborbital: Industry at the Edge of Space*, Springer Praxis Books, DOI 10.1007/978-3-319-03485-0_4, © Springer International Publishing Switzerland 2014

4.1 Experiencing weightlessness on board SpaceShipTwo. Courtesy: The SpaceShip Company/Virgin Galactic

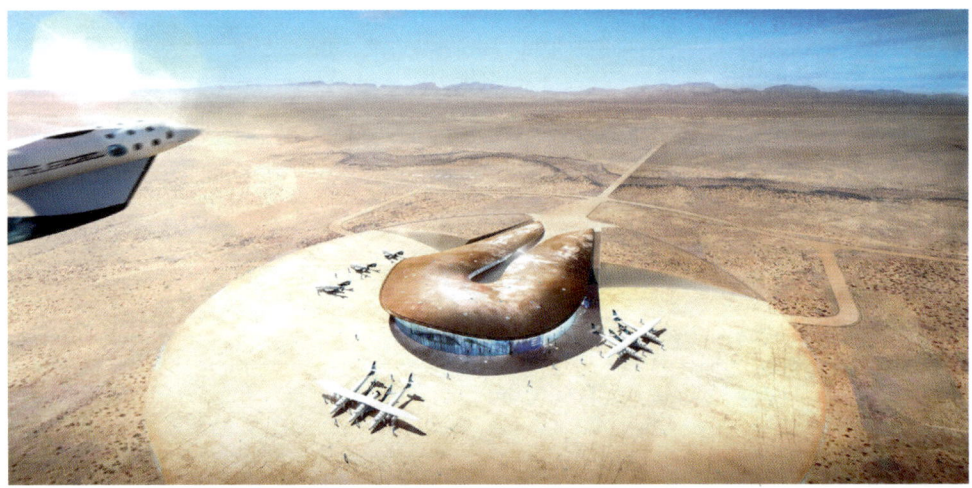

4.2 Spaceport America. Courtesy: Spaceport America

But with the pivotal powered test, revenue flights seemed more tangible, although no date has been set for the first commercial flight. Since the historic X-Prize flight in 2004, more than 600 aspiring spaceflight participants (SFPs) have paid US$250,000 or placed deposits, waiting for a chance to experience weightlessness (Figure 4.1) and view Earth's curvature from a 100-kilometer altitude.

The route the SFPs of tomorrow will take before their flight begins at the Virgin Galactic Gateway to Space (Figure 4.2) – a building whose sinuous steel surfaces stand in harsh contrast to the red-brown desert of New Mexico. Meticulously designed by architects of Foster and Partners to foreshadow the journey the new group of astronauts will make into space, the building features a concrete ramp that ascends gradually towards the center of the structure. Shortly after arriving for their flight of a lifetime, passengers will be issued magnetic tags that will trigger heavy steel doors that will open into a passageway and then a catwalk with views of the cavernous hangar four storeys below, housing the fleet of spacecraft in which they will be launched into space.

Walking along the catwalk, passengers will pass through another set of doors, which will swing open into the astronaut lounge, a vast atrium filled with natural light from an elliptical wall of windows, offering a vista of the three-kilometer-long runway. Today, the talk of spaceports is nothing new. After all, there are dozens of them popping up around the world and nine locations just in the US. But Spaceport America, the New Mexico complex, is the only one built from scratch and designed to accommodate a regular passenger service. It was built from nothing in the middle of nowhere 50 kilometers from the nearest town, and it wasn't cheap; with a price tag of almost a quarter of a billion dollars and counting, the spaceport was paid for by the state of New Mexico, whose citizens voted for a sales tax designed to finance its construction.

Spaceport America is ground zero for the beginning of the commercial passenger space-line industry – the location from where daily suborbital passenger flights will kick-start a whole new era in space travel. For many, it's a dream long overdue. For those who remember the opening scenes of Stanley Kubrick's *2001: A Space Odyssey*, released in 1968, you could be forgiven for thinking that trips to the Moon were just around the corner. Corporate pioneers, Pan American, certainly thought so, and began selling tickets to space, predicting trans-lunar services would begin no later than 2000; 98,000 people signed up. Sadly, the euphoria of *2001* and Apollo 11 didn't last. NASA gutted its lunar program and Pan Am closed its waiting list (the airline went bankrupt in 1991). So began a three-decade-long drought in the annals of commercial spaceflight, until the X-Prize resurrected interest with the pioneering suborbital flights of SpaceShipOne (SS1) in 2004.

It didn't take long after the flights of SS1 before Branson decided to add another sideline to his daredevil brand-building and declare Virgin Galactic open for business. Branson, by then one of the world's richest men and proprietor of his own airline, had long been a fan of manned spaceflight and had even been offered a trip to space in the late 1980s. The offer came from the USSR ambassador in London, who asked the eccentric billionaire if he would like to become the first tourist in space. It would require 18 months' training at Star City and would cost US$50 million. Branson declined. He later regretted his decision, but the offer did spur him on to begin canvassing people about the idea. Then, in 1995, following a discussion with Buzz Aldrin, Branson began seriously exploring the potential for commercializing spaceflight and started to search for a viable space vehicle.

Shortly thereafter, in 1996, the X-Prize, which offered US$10 million to the first team to put a reusable vehicle capable of carrying passengers twice over the threshold of space, kick-started an explosion in the number of private companies claiming they had the technology that could enable the future of space tourism. The leader in the competition was Burt Rutan, whose design reached back to the X-15 rocket planes in which test pilots broke the sound barrier and eventually reached the boundary of space.

4.3   Virgin Galactic logo, featuring Richard Branson's eye. Courtesy: Virgin Galactic

The X-15 was based on a concept study for the National Advisory Committee for Aeronautics (NACA) for a hypersonic research aircraft. Like many X-series aircraft, the X-15 was designed to be shackled beneath the wing of a B-52 bomber (the Balls 8). Release took place at an altitude of about 13.7 kilometers – a height that saved 50 per cent of the fuel it would otherwise have needed if it had been launched from the ground. The X-15 mission architecture is echoed by the one used to launch SS1 on its historic flight and the one SS2 used on its first powered flight in April 2013.

SS1 and its carrier WhiteKnightOne (WK1) were built thanks to more than US$20 million of funding from Microsoft cofounder Paul Allen. Despite the secret-squirrel nature of Rutan's work, word got around that the eccentric aircraft designer was building a spaceship. At the time, Rutan and Branson were collaborating with Steve Fossett, an American businessman and a record-setting aviator, sailor, and adventurer, on the Virgin Atlantic GlobalFlyer, which Fossett flew non-stop around the world in 2005. Branson found out about the spaceship being built in Building 75 and a meeting with Rutan and Paul Allen followed. Rutan and Allen, who had no interest in running a space-tourism company, agreed to license the technology to Virgin and so the Virgin Galactic brand was born. On June 21st, 2004, 64-year-old test pilot Mike Melvill flew SS1 over the Karman Line for the first time. A few weeks later, when Rutan's rocket plane made the two flights necessary to win the X-Prize, the Virgin Galactic logo was on the side of the diminutive spaceship.

At a press conference before the first X-Prize-qualifying flight, Branson announced his intention to launch a passenger service into space. Tickets would go on sale at a cost of US$200,000. Branson, ever the optimist, predicted flights could begin as soon as 2007. A Virgin Galactic website featured the company's distinctive logo (Figure 4.3), footage of the X-Prize-winning flight, and an application form. Not long after the site went live, it crashed due to the volume of requests for tickets.

Seats on the first 100 Virgin Galactic flights were reserved for the first buyers, known as the Founders (William Shatner turned down an offer to be on the inaugural spaceflight, saying "I do want to go up but I need guarantees I'll definitely come back"). This wealthy group would have privileged access to the program as it developed and, when the time came, their names would be entered into a draw to decide who would fly first. Now all that was needed was a spaceship to fly the customers. In Mojave, the Scaled Composites team began work on a vehicle that could meet the requirements of the space-tourist experience. To begin with, the engineers didn't even know how big the spaceship should be. One that carried four passengers? Twelve? Twenty? One thing the engineers agreed on was that Virgin couldn't send its high rollers up in SS1. It was just too small. There was also the business case to consider: to bring the seat price down relatively quickly would mean flying several customers on each flight. After canvassing its customers, spaceship designers started getting an idea of what the typical rich space tourist expected for their investment. Getting out of their seats was a must, as was the ability to see Earth from space; neither of these would have been possible from the confines of SS1. In the end, SS2 was designed to have large windows and to carry two pilots and six passengers. To better understand how to design a cabin for use by space tourists, Rutan's design team took several parabolic zero-G flights in a specially converted Boeing 727.

By summer 2005, Virgin Galactic had banked US$10 million worth of deposit checks and the following year, at the US Wired NextFest, Branson unveiled a SS2 full-size mock-up: a sleek, futuristic white tube with a delta wing, reclining seats molded into soft curves, and a dozen windows. Attending the ceremony were Buzz Aldrin and Alan Watts, a Virgin Atlantic passenger who had saved enough frequent-flyer miles to buy a ticket for space:

> Watts redeemed two million miles for the opportunity to be one of the first space tourists. He had been a member of the Virgin Atlantic flying club for 10 years and also had a Virgin American Express credit card, which awarded him two miles for every pound he spend. The opportunity for him to fly as a space tourist came about when Watts returned home one Friday night and was informed by his daughter that Virgin Atlantic had called and had asked if Watts wanted to go into space. Watts explained he planned to semi-retire within five years and was saving the miles for holidays with his wife, but said he would think about it. On Monday, Watts called Virgin Atlantic and agreed.

Following the unveiling of the SS2 mock-up, Branson announced passenger flights would launch from New Mexico in 2009 - operations are now likely to being sometime in 2014. When complete, SS2 will be much roomier than its predecessor, featuring a cabin 2.28 meters in diameter and 3.6 meters in length (Figure 4.4). That's about half the size of a Cessna Citation X business jet, but riding in SS2 will be very different from riding on board a corporate jet. Slung beneath WK2, its twin-fuselage mother ship, SS2 will slowly ascend to its 15-kilometer launch altitude. The ascent will take about an hour, which will give passengers – and impending astronauts – plenty of time to contemplate their historic journey. Despite the US$250,000 ticket price, there will be no drinks service on board, so the pilots will no doubt talk to their passengers to pass the time and to reassure the anxious.

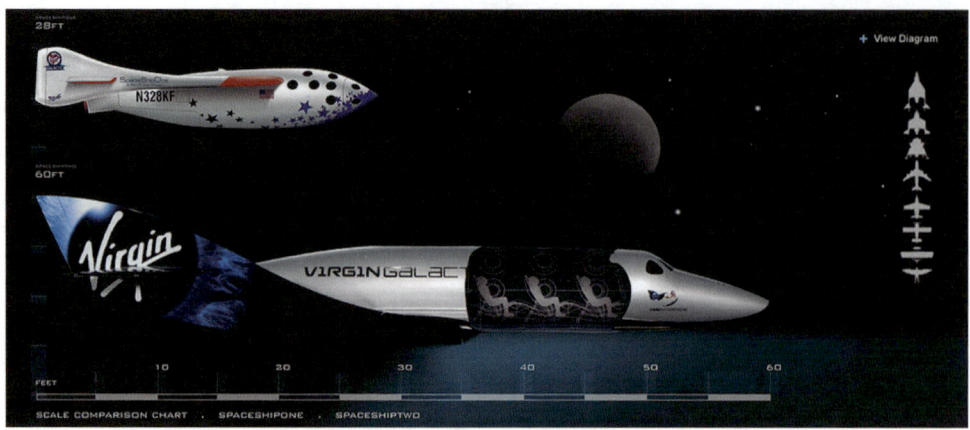

4.4 Technical specifications of SpaceShipTwo. Courtesy: Virgin Galactic

Once released from the WK2, SS2 will fall away to a safe distance, whereupon the pilot will ignite the rocket motor, using two cockpit switches: the first arming the system and the second opening a ball valve, releasing liquid nitrogen dioxide into the throat of the engine. The moment of rocket ignition is guaranteed to get the attention of the passengers as they are thrown back into their seats by the accelerative force. Mach 1 will be reached in just 12 seconds; 18 seconds later, SS2 will be barreling along at Mach 2. One minute into the flight, the spacecraft will be travelling at 4,800 kilometers per hour. Quickly, the sky will turn from blue to navy, to indigo, and then, finally, black. Eighty seconds into flight, the engine will be cut and, shortly thereafter, after releasing their seat belts, passengers – now newly minted astronauts – will get their first taste of zero-G. From their 100-kilometer-plus vantage point, they will be able to see 1,600 kilometers from horizon to horizon, the curvature of Earth, and the blue line of the atmosphere (Figure 4.5). They won't have long to admire the view or to take pictures because SS2 will only spend about four minutes in space.

As the pilot positions the "feather" for-re-entry, the six passengers will fold their seats flat to help them tolerate the 4–5 G of acceleration they will be subjected to during their return to Earth. After a 15-minute glide, they will see the familiar sight of Spaceport America and will be back on the desert runway.

A trip of a lifetime? No doubt. But launching rockets into space is anything but routine; spaceships are not aircraft and, despite Virgin's excellent safety record flying its passengers around the world, and despite Rutan's flawless record in designing radical aircraft, things do go awry. Just ask SS1 test pilot Mike Melvill, who experienced two failures he thought would kill him. Melvill happened to be on site on July 26th, 2007, when a cold-flow test of nitrous oxide went very wrong.

4.5 The view from space. Courtesy: NASA

There were 17 people observing the test, six of whom had taken cover at a mobile command post 130 meters away, where they planned to watch the test on closed-circuit TV. The rest watched from behind a fence, a dozen meters away, as the cold-flow test began. Seconds later, a sudden reaction caused a tank to rupture with such explosive force that the decompressing gas blew 15 centimeters of concrete off the pad beneath the test stand, scattering fragments of rock and carbon fiber. The explosion killed three and injured three more. The California Occupational Safety and Health Administration investigated the accident, noting that Scaled Composites had failed to provide adequate training about the hazards involved with the nitrous oxide rocket fuel the company used in its spacecraft prior to the accident. The investigation also noted that Scaled Composites did not institute a written method or procedure to correct unsafe conditions while conducting the test of the propulsion equipment, nor did it monitor the test site during the time of the accident to ensure employees were not exposed to excess amounts of nitrous oxide. The California State investigation found Scaled guilty of failing to observe correct workplace practices, but was unable to explain what had happened; Scaled launched its own investigation into the accident, calling in experts from Lockheed, Northrop, and Boeing, but nobody could isolate a single cause of the accident.

Rutan stopped work on SS2 and shortly thereafter stepped down from the head of the company he had founded after being hospitalized with heart problems (in April 2011, after 36 years in Mojave, he packed up and left for a ranch in Idaho). Work stopped on SS2 for a year and the company struggled to get back on track. Once again, Virgin Galactic had to revise its forecast for revenue flights from 2009 to 2011, and the estimated costs of the program, first calculated at US$20 million, rose to between US$300 million and US$400 million – at least 15 times the initial estimate. The setback didn't seem to deter potential passengers because the tickets kept selling. At the beginning of 2012, Ashton Kutcher became the 500th person to sign up, joining Stephen Hawking, Philippe Starck, and *Dallas* star Victoria Principal on the list of passengers (incidentally, Virgin doesn't offer complimentary rides, no matter how famous a celebrity might be).

Scaled fixed the problems and, in May 2012, the Federal Aviation Administration (FAA) granted the company an experimental launch permit for SS2. Now, with Virgin Galactic's rocket ship ready for the its next series of powered flights, the final goal is in sight and, if everything goes according to plan, by the time you're reading this, the first suborbital space tourists should have flown. Whenever the first suborbital flight finally happens, Branson says the day he climbs on board SS2 for its inaugural passenger flight will be the most exciting of his life. By launching hundreds and eventually thousands of passengers into space, he hopes to give birth to a new industry and, with that in mind, the Spaceship Company has already begun construction of a second spaceplane and mother ship.

XCOR

In the years since Branson's Virgin Galactic brand first entered the fledgling business of blasting tourists to the edge of space, commercial spaceflight – New Space – has become more and more crowded. And competitive. One of Virgin's competitors is XCOR Aerospace, which will use its Lynx (Figure 4.6) to ferry its passengers on their suborbital ride.

Since its founding in 1999, the small, Mojave, California-based company has built a solid reputation for steady and incremental progress. The company has successfully built rockets and rocket engines before and, in many ways, the Lynx is seen as another step on a technology path towards competing in the space-tourism marketplace. Andrew Nelson is XCOR's Chief Operating Officer and Vice President of Business Development. He's responsible for leading XCOR's business team that deals with establishing commercial operations of the Lynx at operating locations around the US and abroad, regulatory compliance and export licensing, sales and marketing functions of the company, and intellectual property strategy. As well as being a recognized leader in New Space, Nelson is the originator of the Space Vehicle Wet Lease concept that is at the core of XCOR's market strategy. The concept allows sovereign countries, corporate entities, and individuals the opportunity to experience the benefit of their own manned spaceflight program without the headaches associated with operating and maintaining a spaceship. An example of the application of the wet lease concept is XCOR's Memorandum of Understanding (MOU) for the wet lease (pending US government approvals) of a production version of the Lynx to be stationed on the island of Curaçao in the Netherlands Antilles. The MOU, announced on October 5th, 2010, had a planned start date in January 2014, when Space Xperience Curaçao (SXC) was to begin marketing, and XCOR was to begin operating, suborbital space-tourism flights and scientific research missions out of Space Port Curaçao (Figure 4.7).

The MOU came about following the Curaçao government and airport authority announcement of their intentions of investigating and creating the conditions suitable for the formation of a commercial spaceflight services industry. SXC's ambition is to create a

4.6   The author in the passenger seat of the Lynx mock-up. Courtesy: Jason Reimuller

4.7   Spaceport Curaçao. Courtesy: www.caribbeanspaceport.com/www.spacexc.com

major tourist attraction for the Caribbean, while offering a venue for international scientific space research. Its vehicle of choice is the Lynx, chosen thanks to its innovative but straightforward and robust design, as well as its enormous commercial potential and competitive viability.

In addition to brokering wet lease agreements, Nelson has been responsible for the successful fundraising and business development program at XCOR that has resulted in significant investment and revenue for the company. He has also led the company's efforts in building the engine development and sales business at XCOR that has produced aerospace supplier clients such as United Launch Alliance.

The Lynx will take off from a runway, just like an aircraft, and will climb just as high as SS2, where the sole passenger will be able to view Earth's curvature and experience four minutes of weightlessness, although he/she won't have as much room as their Virgin Galactic counterparts because they will sit in the co-pilot's seat, which means flying around the cabin is a non-starter. The entire flight will take about 25 minutes and passengers can expect to pay US$95,000 for their flight.

With so much attention directed on the spectacular efforts of SpaceX in the world of orbital operations and the media extravaganza that is Virgin Galactic, it's sometimes been difficult not to lose track of the Lynx. But, in the same Mojave Desert airfield where Scaled Composites is developing SS2, XCOR's vehicle is closing in on the launch of revenue flights. Although the sporty spaceship only has two seats, its low weight and high octane fuel give it some important advantages, including direct runway launches without the complication of a mother ship and the ability to fly several times per day, which should equate to cheaper flights. Like SS2, the Lynx is rocket-powered, but that's about all the vehicles have in common. To begin with, the diminutive Lynx, with its 190-knots take-off speed, gets off the line (it can get airborne with only 400 meters of runway) a whole lot quicker than SS2.

Powered by four kerosene and liquid-oxygen engines, the Lynx's all-liquid design is more efficient than SS2's hybrid propulsion, providing more thrust per pound of fuel; the all-liquid fuel should also give it faster turnaround between flights because all crews will need to do will be to top off the tanks and go again, whereas SS2's engine has to be replaced between flights. Thanks to this quick turnaround, which is expected to take no longer than two hours, XCOR reckons they can fly four missions a day. Then there is the safety issue. Think about having a team of mechanics poring over an engine as two-thirds of it is being replaced after every flight, as is the case with SS2, and compare that scenario with the Lynx, which is effectively a hands-off operation. In some people's minds, the Lynx is the safer operation.

The first Lynx iteration is designated the Mark I, a prototype, designed to reach two times the speed of sound and a maximum altitude of 60 kilometers, which isn't space, but still allows a couple of minutes of weightlessness. The planned ceiling of the production model Lynx, the Mark II, is 106 kilometers. On the technical side, the Mark I will feature a carbon-composite skin that will be modified as the flight test program progresses and will allow XCOR to adjust the vehicle's aerodynamics. In contrast, the Mark II will feature a special composite that will better withstand the heating caused by re-entry. Looking towards the horizon, there is the Mark III, which will feature an enclosed bay on the top, from which a separate rocket booster will launch a small satellite.

The experience of flying on board the Lynx will be a little different from flying on SS2. For one thing, there are the more cozy confines and, for another, passengers won't be allowed to unstrap after engine cut-off. Both pilot and passenger will wear pressure suits as a safety measure in case cabin pressure is lost during the flight.

Lynx Step by Step

Lynx has an all-composite airframe and a thermal protection system (TPS) on the nose and leading edges to deal with the heat of re-entry. The double-delta wing area is sized for landing at moderate touchdown speeds near 90 knots. Measuring nine meters in length with wings that span 7.5 meters, the Lynx is in the sports car category of spacecraft.

The Lynx Mark I

The Lynx Mark I is a prototype vehicle that will be used to characterize and flight test the vehicle's subsystems including life support, propulsion, tanks, structure, aeroshell, aero-dynamics, and re-entry heating. Designed to reach an altitude of 60 kilometers, the vehicle will be used to train pilots and crew for the Lynx Mark II.

The Lynx Mark II

The Mark II is the production version, designed to service the suborbital tourism market and other markets that make use of the vehicle's payload volume. The Mark II, which is designed to reach an altitude of 100 kilometers, uses the same propulsion and avionics systems as the Lynx Mark I, but has a lower dry weight and hence higher performance.

The Lynx Mark III

The Lynx Mark III is a modified version of the Lynx Mark II that features an external dorsal pod capable of carrying a payload experiment or an upper stage capable of launching a small satellite into low earth orbit (LEO). The Mark III features upgraded landing gear, aerodynamics, core structural enhancements, and a more powerful propulsion package than the Mark II.

Propulsion

Four XR-5K18 rocket engines, each producing 12.9 kN (2,900 lbf) vacuum thrust with kerosene and liquid-oxygen propellants, provide the power to launch the Lynx into space. The engine, which features XCOR's proprietary spark torch ignition system, has the ability to stop and restart.

Payload Mission Capabilities

The Lynx will offer a variety of multi-mission primary and secondary payload capabilities ranging from in-cockpit experiments and externally mounted experiments to astronaut training and personal spaceflight. Lynx vehicles will carry their payloads in the area to the right of the pilot or, in the case of the Mark III, in an experiment pod. For the Mark II version, the primary internal payload will accommodate a maximum mass of 120 kilograms to 100 kilometers, while the Mark III vehicle will be capable of carrying up to 650 kilograms in its external dorsal mounted pod – large enough to hold a space telescope or a carrier to launch multiple nanosatellites into LEO.

5

Contenders: Vehicles Waiting in the Wings

Copenhagen Suborbitals' mission patch for the Tycho Brahe. Courtesy: Copenhagen Suborbitals

While Virgin Galactic and XCOR have taken the lion's share of media coverage in the suborbital space race, these aren't the only companies with their sights set on New Space. Several other companies are hard at work designing, building, and testing vehicles that also aim to provide rides to suborbital space and, in this chapter, we profile four of them: Blue Origin, Armadillo Aerospace, Masten Space Systems, and Copenhagen Suborbitals.

E. Seedhouse, *Suborbital: Industry at the Edge of Space*, Springer Praxis Books, DOI 10.1007/978-3-319-03485-0_5, © Springer International Publishing Switzerland 2014

BLUE ORIGIN

Blue Origin is an entrepreneurial space company working on a cone-shaped space vehicle (Figure 5.1) designed for suborbital and (eventually) orbital trips. Although the company has a reputation for being stingy releasing information, the curtain of secrecy is gradually being raised.

At the company's helm is Jeff Bezos, who made his fortune as the founder and CEO of Amazon and who was named *Time*'s Person of the Year in 1999. According to Forbes, Bezos had an estimated net worth of US$23.2 billion as of September 2012. Even as a teenager, Bezos wanted to get involved in space, but as an entrepreneur rather than as an astronaut. To achieve such a heady dream, Bezos reckoned he would need to make a lot of money. An awful lot of money. He chose the field of computer science to make his fortune and, in 1994, founded Amazon with a plan to sell books over the Internet. From modest beginnings in his garage, Bezos expanded his Amazon brand to an empire selling everything from books and baby products to tools and toasters. Almost two decades later, Bezos is still innovating the Amazon brand, leading the charge with the e-publishing movement. But the industry where he is making the biggest waves is space.

As of early 2013, the Kent, Washington-based company has completed wind tunnel testing of its spacecraft, simply called the Space Vehicle, which will be capable of

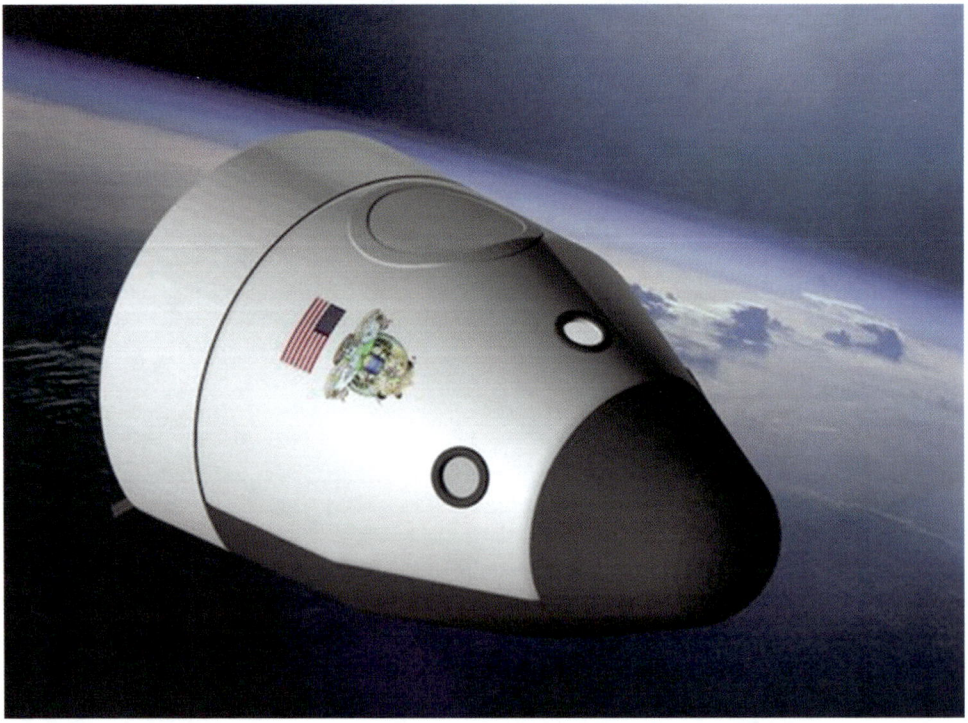

5.1 Blue Origin's New Shepard spacecraft. Courtesy: Blue Origin

transporting up to seven astronauts to low Earth orbit (LEO) and the International Space Station (ISS). The Space Vehicle's biconic shape is the product of a heavy dose of computational fluid dynamics (CFD) analysis that has been validated by hundreds of hours of testing at Lockheed Martin's High Speed Wind Tunnel Facility in Dallas. The testing was conducted as part of Blue Origin's partnership with NASA, under the agency's Commercial Crew Development (CCDev) program, which awarded the company US$22 million in 2011 to develop the vehicle. Under CCDev, Blue Origin is prepared to start tests of its BE-3 engine, which will be used on Blue Origin's reusable launch vehicle. Other elements being tested by the company include a "pusher" launch-abort system and evaluation of an ability to control the flight path of a subscale crew capsule using a thrust vector control system.

Founded in 2000, the company first made headlines in 2006 when Bezos made a series of land purchases in Texas. According to the *Wall Street Journal*, the purchases were made under corporate names (all based on famous explorers) such as "James Cook L.P.", "Jolliet Holdings", and "Cabot Enterprises", which all traced back to the same address. The following year, Blue Origin's website opened for business, featuring videos and photos of such milestones as the November 2006 test flight of its New Shepard vehicle. In common with some of the other companies in the New Space industry, Blue Origin has had a helping hand from NASA, having received two rounds of funding from the agency: US$3.7 million in 2010 for the first round of the Commercial Crew Program (CPP) and US$22 million for the second round in 2011. Until quite recently, it was difficult to know how well or how badly development of their spacecraft was going, although the company did disclose a major setback in 2011, when it lost a development vehicle due to flight instability during a flight test. "A flight instability drove an angle of attack that triggered our range safety system to terminate thrust on the vehicle. Not the outcome any of us wanted, but we're signed up for this to be hard," wrote Bezos in a September 2nd, 2011, update on the company's website. A year later, the news was more positive with the successful tests of the company's crew capsule escape system, which ascended to 703 meters before returning by parachute, demonstrating a key safety system for suborbital and orbital flights.

In the short term, the company is developing an early prototype suborbital vehicle that will carry three or more people into space from a launch site in Texas. By starting with suborbital flights, the company hopes to gain flight experience that can be used to lead develop orbital spacecraft.

NASA is hoping Blue Origin can translate that experience sooner rather than later so they don't have to pay through the nose for their astronauts to be ferried to the ISS on the aging Soyuz; the latest cost for a seat on the Soyuz is more than US$70 million. But, before Blue Origin can transport NASA astronauts to the ISS, it must first achieve success in the suborbital arena, and that means developing Goddard, their suborbital test vehicle.

Many compare the Goddard, named after rocket pioneer Robert Goddard, to a NASA test vehicle, the Delta Clipper (or Delta Clipper Experimental – DC-X; Figure 5.2), because of its appearance and launch style. The DC-X was an unmanned prototype of a reusable single stage to orbit (SSTO) vehicle built in conjunction with the US Department of Defense's Strategic Defense Initiative Organization (SDIO) from 1991 to 1993. Between 1994 and 1995, testing continued through funding of NASA and, in 1996, the DC-X technology was transferred to NASA. The elegant DC-X was never designed to

5.2   Delta Clipper. Courtesy: NASA

achieve orbit, but to demonstrate the concept of vertical take-off and landing (VTOL). The design used attitude control thrusters and retro rockets to control the descent, allowing the vehicle to begin re-entry nose-first, before rolling around and touching down on landing struts at its base. The vehicle, which needed a crew of only three to man its control center, could be refueled where it landed and take off again from the same location – a feature that allowed very fast turnaround times.

The DC-X's aeroshell was custom-constructed by Scaled Composites, the same company that designed SpaceShipOne (SS1), with most of the rest of the spacecraft being built from off-the-shelf components. The DC-X first flew for 59 seconds on August 18th, 1993, and on two more occasions before funding ran out. More funding was provided by NASA and the Advanced Research Projects Agency and the program restarted on June 20th, 1994. Testing continued until July 7th, 1995, when a hard landing cracked the aeroshell. By this point, funding had been cut and NASA adopted the program and applied a series of upgrades to test new technologies. The upgraded vehicle was called the DC-XA and resumed flight in 1996. Unfortunately, flight testing didn't go so well and, on July 7th, 1996, a cracked liquid-oxygen tank caused such extensive damage that repairs were deemed impractical. Instead, NASA focused development on the Lockheed Martin VentureStar and several engineers who had worked on the DC-X were eventually hired by Blue Origin.

The Goddard, with its cone-shaped nose and blunt base, launches and lands in a vertical position and sits on four legs. As a test vehicle, it will be flown to progressively higher altitudes to test the viability of the vehicle that will eventually ferry tourists to space: New Shepard. Named after the first US astronaut in space, Alan Shepard, New Shepard will be a reusable launch vehicle (RLV) capable of ferrying three or more passengers to 100 kilometers. Propelled by rocket-grade kerosene and high test peroxide (HTP) rockets, New Shepard flights will last less than 10 minutes (compared to 90 minutes for SpaceShipTwo flights and 25 minutes for XCOR Lynx missions) from start to finish. In one mission architecture, New Shepard's engines will shut down after two minutes, allowing it to coast to suborbital altitude. Then, for re-entry, the engines will restart, allowing the vehicle to land from where it took off. In an alternative mission architecture, the crew capsule will separate from the rocket propulsion module during flight. The crew capsule will then land by taking advantage of atmospheric drag, perhaps with parachutes. Whichever architecture is used, Blue Origin will likely have the most elegant mission operation, made leaner by the fact New Shepard will work autonomously – without *any* ground control. Instead, the vehicle has on-board systems allowing the crew to control it, although a ground operator will be on hand to assist passengers in the event of a contingency.

Speaking of contingencies, it's worth taking a closer look at Blue Origin's emergency crew-escape system, which is based on a pusher motor, rather than the conventional tractor system employed by the Soyuz. Each type of system is designed to carry the crew capsule away from the launch vehicle in the event of an emergency during launch or during the early stages of ascent. The problem with the tractor system is that, following a successful launch, the tower on top of the capsule is jettisoned after a specific altitude is reached. The advantage of the pusher configuration is that it provides a crew-escape capability without the need to jettison the unused escape system; it's a configuration that will give passengers peace of mind and also provide another level of safety in what has always been a high-risk venture.

Unlike some companies that rely heavily on financial assistance, Blue Origin intends to go forward with or without the space agency, although they acknowledge NASA's commercial crew office has helped accelerate their plans. While the company might stand apart from some of the more colorful personalities found in other New Space companies, there is no doubt Bezos's operation is an intense engineering, technical company involved in a long-term effort that it's pursuing incrementally, step by step. Blue Origin may not have reams of information on their website (much of what is on the website was released as a result of mandatory disclosures to the Federal Aviation Administration (FAA) and NASA as Bezos sought regulatory approvals and funding), but this New Space company is more about talking about things they've done and not things they're planning to do. It's more about accomplishments – an approach that allows the company to not set any false expectations and instead maintain focus on the development of their space vehicle and developing the capability to be responsive and cost-effective.

ARMADILLO AEROSPACE

A little further back in the commercial space race is Armadillo Aerospace (Figure 5.3), one of the original 19 space start-ups that vied to win the US$10 million Ansari X-Prize. Funded largely by video game developer John Carmack, of *Doom* and *Quake* fame, the company has taken major strides with its FAA-approved suborbital rocket, the STIG-B (named after the anonymous driver that features in the BBC television show *Top Gear*; Figure 5.4). Despite encountering difficulties reaching design altitude during testing, the company appears to have mastered the landing side of the equation using a self-guided ram air parachute system: a flight on December 6th, 2012, returned the rocket to within 55 meters of the intended recovery point.

The STIG-B is rocket is 0.5 meters in diameter and 10 meters long (STIG-A was 0.4 meters in diameter and 9 meters long) and is designed to lift 50 kilograms of payload to an altitude in excess of 100 kilometers. The rocket is designed as a test bed for technology to be used on a manned suborbital vehicle and Armadillo Aerospace's goal is to fly the rockets once per month for a total of 24 launches. As you can see in the photo, the STIG-B lands on the nozzle edge of the rocket engine. While this is an unusual choice, the rocket engine happens to be one of the strongest structures on the fuselage. Among other features is the recovery system comprising a ballute (balloon-parachute) that is ejected from the nose of the rocket. There is also an APRS (Automatic Packet Reporting System) mounted in the nose cone, which relays GPS locations and allows the nose cone to be easily tracked for recovery using a Smartphone GPS locator app.

The third STIG-B launch was conducted in February 2013 in Las Cruces, New Mexico. Given the scale of the STIG-B, the logistics were fairly straightforward; the team simply

5.3 Armadillo Aerospace. Courtesy: Armadillo Aerospace

5.4   Armadillo's STIG-B. Courtesy: Armadillo Aerospace

parked the launch control trailer and the crane truck station beside the launch pad, began unpacking the trailer, and prepared for a dry-run launch. Because of the cold, the team bought electric blankets from Wal-Mart to keep the vehicle's batteries warm; the rest of the launch operation equipment was kept running thanks to a diesel generator linked to an auxiliary diesel tank so it could run overnight at full power without needing to refuel. Not exactly a glitzy Virgin Galactic setting! After trouble-shooting some GPS issues (it took some time for the GPS to acquire a lock) and verifying the GPS was operational, it was too late in the day to begin a dry run so the team decided to delay until the next day. The following day, the Armadillo team simply wheeled the vehicle on its trailers with a four-wheeler and conducted a successful dry run. Weather on launch-day morning was fog and low cloud cover, which caused a problem because a 1,500-meter ceiling and 10-nautical-mile visibility was required. With the noon launch window approaching, the weather finally began to clear and the team loaded the propellants. The vehicle cleared the launch stand and all seemed to be proceeding according to the script. Unfortunately, the GPS signal strength dropped as soon as the rocket launched – an issue that hadn't been observed in the tie-down, hot-fire test. More problems began when the vehicle reached a velocity of 70 meters per second and the team lost the signal altogether, although it was reacquired when the velocity dropped below 70 meters per second. The pitch and yaw angles appeared to be well controlled resulting in a good attitude for deploying the ballute and nose cone. Indications were good for a successful recovery of the vehicle until an anomalous release caused by a hang-up in the recovery system prevented the main from pulling out of the deployment bag. The result was a debris field 700 meters from the launch pad. Because of the loss of vehicle, the team had to file a mishap report with the FAA/AST representative, which resulted in a recommendation that the recovery system be resolved (the GPS wasn't required to be addressed because it wasn't considered part of the flight safety system).

While it wasn't the result the Armadillo Aerospace team were hoping for, there were many positives they took away. The STIG-B had flown three times in almost as many months, which was the first time a licensed, fully reusable, liquid-bi-propellant rocket had been flown that many times. Also, despite loss of the vehicle, Armadillo recovered their payloads intact, and the team learned a lot about the payload integration process. The company had also obtained their first full license from FAA/AST, and had attracted the attention of several government-funded and private commercial payload providers.

Armadillo's game plan moving forward is to build a fleet of three or more STIG vehicles to allow the company to campaign the STIG vehicle for commercial scientific payloads and even suffer some damage without having to halt the campaign for too long. While the STIG flights are unmanned, the STIG vehicle is a precursor to a more ambitious, vertical take-off/vertical landing vehicle, the Black Armadillo, which will have room for a pilot and passengers.[1] Details of Black Armadillo's progress are sketchy but Armadillo already has an agreement with tour operator Space Adventures of Vienna, Virginia, to market trips on the vehicle. While Armadillo passengers wait for their opportunity to fly on Black Armadillo, the STIG will continue to provide Armadillo with experience of real space conditions, including checks to ensure all systems continue to work in space and that the team on Earth can maintain communications with and recover the vehicle. All this information will be transferable to Black Armadillo, which will use many identical systems but have a larger airframe and eight engines rather than one.

MASTEN SPACE SYSTEMS

Although the lion's share of media coverage in the suborbital space industry has focused on those companies planning to launch manned flights, there are a few companies working on unmanned vehicles. Leading the charge is Masten Space Systems. Working from Mojave, California, Masten is leading the quiet unmanned revolution by developing new reusable VTOL rockets. Although VTOL vehicles are perhaps the most elegant way to travel to space, the actual realization of such a vehicle is a surprisingly difficult engineering challenge. But, step by step, aerospace start-ups like Masten are working the problem from the ground up, flying a little higher with each test flight and each iteration of their rockets. And, for the most part, they have been very successful. In October 2009, Masten's Xombie (Figure 5.5) won the US$150,000 second prize in the Level One competition of the Lunar Lander Challenge with an average landing accuracy of 16 centimeters,

---

[1]   Armadillo Aerospace announced in October 2010 that a Russian had become its first confirmed passenger. St. Petersburg resident Evgeny Kovalev won his ticket to space in a contest organized by Efes brewery. Currently, there are some 200 applicants for the flights, which will cost about US$100,000.

5.5 Xombie. Courtesy: Armadillo Aerospace

demonstrating stable, controlled flight using a guidance, navigation, and control (GN&C) system developed in-house.

Xombie was followed by Xoie, an aluminum-framed vehicle that won the company the US$1,000,000 Level Two prize of the Lunar Lander Challenge, beating fellow aerospace startup Armadillo Aerospace. Next in Masten's mini VTOL fleet was the Xaero, developed in 2010–11 as a potential vehicle for use as a suborbital reusable launch vehicle (sRLV) for carrying research payloads under NASA's Flight Opportunities Program. The first Xaero test vehicle flew 110 test flights before being destroyed in its 111th flight. Next up was XEUS (pronounced Zeus), a VTVL lunar lander demonstrator comprising a Centaur upper stage (from United Launch Alliance) and a RL-10 main engine to which four vertical thrusters have been added. Production XEUS vehicles are estimated to be able to land on the Moon with up to 14 tonnes of payload when using the expendable version or five tonnes of payload when using the reusable version.

Masten's progress received a boost in 2011 when it received multiple contracts from NASA's Flight Opportunities Program, which pays for flights aboard reusable suborbital vehicles (one of which is Masten's Xaero) for qualified research experiments. If you happen to be a cash-strapped researcher or scientist who wants to hitch a ride on a suborbital rocket, then Masten is the company for you. The first thing you need to do is download a copy of their Payload User's Guide (see Appendix III), which provides information on the capabilities of Masten's spacecraft and the environment expected aboard their vehicles. If you'd like to speak to someone about how your payload might be flown, the company provides a phone number to call (678-551-2253).

Early in 2012, Masten continued its work developing its precision landing system. Some may wonder why this is so important to the company; after all, shouldn't their attention be focused on developing its group of high-altitude suborbital vehicles? The answer Masten provides on its website is that precision landing addresses a major market need – in fact, in the National Research Council's Space Technology Roadmaps and Priorities report, precision landing was named one of the top technical challenges facing the US space program. Importantly for Masten, developing a precision landing system also allows the company to contribute to the national space program in a tangible manner as well as to mature a rocket-recovery mechanism that will enable higher-tempo launch operations.

In addition to working on its precision landing system, 2012 saw Masten testing new engines. The company has designed all sorts of classes of engines, the biggest of which (up to early 2012) has reached 1,200 pounds of thrust. The newest prototype to be tested in 2012 was the Katana-class KA5S, which produced about 2,000 pounds of thrust. The test boded well for Masten's suborbital ambitions because the Katana-class engines will eventually produce up to 4,000 pounds of thrust and will power the company's Xogdor-class suborbital vehicle as well as its XEUS lander demonstrator. Later in the year, Masten continued testing of its Xaero, which culminated in a flight to 444-meter altitude, the highest any Masten vehicle had flown to date. Shortly after the Xaero flight, the company completed the third stage of its Xombie flight campaign, which pushed Xombie to a new performance envelope. The flight ascended to 476.4 meters before translating downrange 750 meters at a horizontal velocity of 24 meters per second. All flight test objectives were reached and the flight once again underlined Masten's precision landing capability.

Like all companies involved in the testing of rocket engines, failures are part of the game, and Masten has had its fair share of them. In September 2012, a month after the Xombie flight, Masten's Xaero vehicle encountered a subsystem failure in flight that necessitated turning the rocket off while it was still in the air. While it is never easy to lose a test vehicle, flying the Xaero was an iterative step for Masten, and they bounced back quickly by developing the Xaero B that was fitted with a new power-plant: the Scimitar engine. Known for using lightweight all-aluminum engines, the Scimitar boasted a bare-engine thrust-to-weight ratio of 133:1, an engine-module thrust-to-weight ratio of 40:1, and an impressive Isp efficiency of 94%. If you happen to be a rocket engineer, those numbers speak for themselves and, for those who aren't, suffice to say the Scimitar has a lot of punch.

In early 2013, Masten announced they were offering an open loop flight on their Xombie VTVL rocket for a sensor or experiment. Researchers or engineers could choose from a 60-second hop up to 500-meter altitude or a translation flight 51 meters downrange across the Mojave desert. The US$99,563 opportunity included luxury accommodations at the Mojave Days Inn, breakfast with Masten founder and CTO Dave Masten, and a tour of Masten's production facility. Another sign the unmanned suborbital flight industry was gathering steam was Masten's announcement that they had begun taking payload slot orders. Payloads are sold in one-kilogram increments at US$250 per kilo so, if your payload happens to weigh five kilograms, you would pay US$1,250. For those who can't afford the custom payload package, Masten offers CanSat flights for US$99. Not a bad price to get your payload into suborbital space!

COPENHAGEN SUBORBITALS

Kristian von Bengtson designs and builds spacecraft. Nothing unusual about that, but Copenhagen Suborbitals (CS), a company he founded in May 2008, is not your run-of-the-mill spaceship company. For one thing, it's open-source and, for another, it's a non-profit enterprise. Five years after being founded, the company has thousands of donors, several sponsoring companies, and more than 40 full- and part-time volunteers. In 2011, the company succeeded in launching their hand-built dummy manned rocket (Figure 5.6) for just US$100,000. While the vehicle encountered a trajectory anomaly, the CS team were still able to communicate with the rocket, shut down its hybrid rocket engine, separate the spacecraft, and deploy the parachutes. Not a bad achievement for the cost of what NASA might spend preparing a PowerPoint presentation. Here's their story.

From the outset, CS was intended to be a space endeavor, based entirely on private donors, sponsors, and volunteers. Their mission is a simple one: to launch human beings into space on privately built rockets and, by doing so, show the world that manned spaceflight doesn't have to be expensive. The project is open-source and non-profit to inspire as many people as possible, and to involve CS's partners and their expertise. Work takes place in a 300-square-meter storage building called the Horizontal Assembly Building (HAB), situated on an abandoned shipyard in Copenhagen, a location that provides CS with plenty of room to test their rocket engines (Figure 5.7). With no administration or technical boards to approve their work, this company moves *very* fast from concept to bending metal.

5.6 The confines of Copenhagen Suborbitals' Tycho Brahe isn't a place for claustrophobes.
Courtesy: Copenhagen Suborbitals

5.7 The Horizontal Assembly Building (HAB) – Copenhagen Suborbitals' version of the
Vehicle Assembly Building – just a little scaled down. Courtesy: Copenhagen Suborbitals

Leading the team are Peter Madsen, the company's launch vehicle lead, and Kristian von Bengtson, who takes care of flight director duties. Helping them are a small platoon of volunteers, including space medicine specialist Niels Foldager, physicist Steen Eiler Jørgensen, Bo Brændstrup, who leads the development of the CS space capsule, and spacesuit developer, Cameron Smith. One of the principles that CS has adhered to is to stay away from the use of toxic, corrosive, or dangerous propellants. All propellants – especially oxidizers – are dangerous if not handled properly, but CS deals with those with the fewest problems, which means they use LOX extensively.

To get a better understanding of hybrid propulsion, CS used a throttleable engine originally intended to propel a boat – the XLR-2. To boost the engine's performance, CS went to work combining mixes of epoxy resin and nitrous oxide that increased the engine's power from 1 to 4.5 kN. Next in the rocket engine learning curve was the HATV, a nitrous oxide/epoxy motor capable of producing a thrust of 12 kN, with a combustion time of 20 seconds. CS's experience testing the HATV represented a key step that bridged the gap from amateur-sized rocket engines to the larger, more powerful ones that will eventually carry a crew.

After the HATV came the HEAT-1X, a 64-centimeter-diameter polyurethane rocket that was built in the spring of 2010. Intended to fly a full-scale mission with the Tycho Brahe spacecraft on top, the HEAT-1X underwent static testing later that spring, the results of which boded well for a launch attempt later in the year. Unfortunately, the launch was aborted due to a LOX valve failure. The LOX valve wasn't the only headache the CS team faced. During the development of their LOX-based hybrids, the engines suffered from persistent moderate combustion efficiency. To resolve the problems, they tried experimenting with turbulence-inducing baffles in the combustion chamber, but this measure didn't resolve the issue. So, to gain a better understanding of the problem, CS conducted 12 tests of a bi-fuel liquid-rocket motor, eight of which resulted in fine burns and surviving engines. And, in preparation for testing bigger 100-kN engines, they built a bigger test stand. In parallel with this work, a modified HATV – Sapphire – is being readied for launch and development continues on the space capsule, which involves designing parachutes, life-support systems, heat shield, aerodynamics, cabin-pressure control, and ground communication. Which brings us to the Tycho Brahe II (Figure 5.8).

Named after the Danish astronomer, the Tycho Brahe is designed to support one person on a suborbital flight (Table 5.1). Designed by Kristian von Bengtson, the space capsule will be flown in a series of test flights carrying a 50th-percentile 70-kilogram dummy.

As you can see in Figure 5.8, the Tycho Brahe is not for claustrophobes. The capsule is launched with the crewmember in the standing orientation, which dramatically reduces the rocket diameter and also provides a very good view. The only problem – apart from claustrophobia – is the astronaut's orientation with respect to G-loads (+Gz in this case). CS say their calculations indicate the G-load time won't be a problem when combined with their seating system, which will allow the astronaut to slightly bend their legs and provide support for their neck and head. Of course, even with the seating system, the astronaut won't be able to move around, although he/she will have limited use of their arms to perform basic operations. After its trip to suborbital space, the capsule's pressure hull will be protected from the heat of re-entry thanks to a 15-millimeter heat shield made from one-millimeter cork-layers, after which the capsule will descend under three parachutes, which

5.8   Test flight of the Tycho Brahe. Courtesy: Copenhagen Suborbitals

Table 5.1. Tycho Brahe.

Mass	300 kg (including astronaut)
Length	3.5 m
Diameter	64 cm
Internal atmosphere	1 bar
Pressurized volume	650 L
Unpressurized volume	200 L
Parachutes	4 (1 drogue, 3 main)
Heat shield	15-mm cork
Communication	2 downlink, 1 uplink
Tracking	10.5-Ghz transponder antenna
Power	12/24 V DC
Emergency egress	Personal parachute

are still undergoing development. The capsule has already undergone a number of tests. On June 3rd, 2011, the Tycho Brahe was launched using the HEAT-1X engine and flew for about two minutes, before being recovered in the Baltic Sea. Due to a trajectory anomaly, the parachutes had to be deployed at great speed to save the capsule. This flight test was followed by a pad-abort test, which was completed on August 11th, 2012.

Over the span of a few years, CS has developed a launch vehicle and crew capsule for one brave astronaut, undertaken more than 30 static engine tests, and raised all the money from donors and sponsors to keep the launch attempts on track. Unlike Blue Origin or Virgin Galactic, CS is not a business, nor is it an attempt to race to be the first to achieve manned suborbital spaceflight in Europe. CS is all about pushing the limits of a small group of individuals and, in that light, it has to be applauded. In an industry where it seems only governments and companies with hundreds of millions of dollars can succeed, CS reminds us that an unconventional group of rocketeers can find the pioneering spirit, exhibit true entrepreneurship, and, hopefully along the way, inspire as many as possible. Von Bengtson's and Madsen's ultimate goal is a rocket ride to suborbital space. That may be some time away but you have to admit the CS concept is intriguing. I for one hope they achieve their goal.

6

Spaceports

The ink was barely dry on the US$10 million Ansari X-Prize-winning check before a potentially equally lucrative space race was announced: the competition between spaceports. Kick-starting the contest was Peter Mitchell, director of the New Mexico Office for Space Commercialization, who was present at Mojave Airport on the day that SpaceShipOne (SS1) made history. "Today doesn't belong to New Mexico, the day belongs to the gentlemen up here … that made this dream become a reality," Mitchell told reporters. "Tomorrow, however, we focus on bringing the spoils of this dream to the state of New Mexico." Not surprisingly, the remark rubbed some people the wrong way, including Dick Rutan, brother of SS1 designer Burt Rutan, and a member of the Mojave Airport district's board of directors. He promised to give Mitchell a run for his money and the race was on.

Remember, this was back in 2004, before the suborbital passenger business was even a business. But, with market studies predicting suborbital space tourism could generate more than a billion dollars a year in revenues by the year 2021, the nascent industry's main players reckoned space passenger operations would require upgrades to attract the first wave of deep-pocketed thrill-seekers. After all, these passengers would need a place to train and a suitably high-end resort to stay. Not the sort of facilities you normally find at rocket-launch ranges, which generally have lots of wide-open space but little else. In short, there needed to be a viable tourist destination, and so the spaceport was born.

What is a spaceport? Perhaps one way to answer that question is to consider the types of people who make up the suborbital space industry and to review the features you can expect when you visit one of these facilities. We know the industry will depend upon two distinct kinds of space adventurer. First, there will be the wealthy few who can afford to buy a ticket and the commercial astronauts employed to conduct research or fly a payload. That's the first category. Those in the second category are neither rich, nor do they have aspirations to be commercial astronauts, but they hope to be able to afford the experience of a flight when prices come down. This second category will come to spaceports to witness the experience and will want to feel vicariously involved in the spaceflights. They will spend money on accommodation and food and drink, and on souvenirs. So the most important thing is that they are able to get there. And, when they arrive, they will want to

E. Seedhouse, *Suborbital: Industry at the Edge of Space*, Springer Praxis Books, DOI 10.1007/978-3-319-03485-0_6, © Springer International Publishing Switzerland 2014

6.1   Dunker training: necessary training for future spaceflight participants. Courtesy: A4H

feel relaxed and welcome. An environment similar to a commercial airport or a cruise ship terminal will be needed. Each category will obviously want to tell their friends about their spaceflight experience, whether real or vicariously experienced, so easy communication with the outside world will be expected.

Next are the training facilities. Future spaceflight participants will need training facilities, which will be co-located at the spaceports. Centrifuges, hypobaric chambers, spatial disorientation trainers, classrooms, dunker training equipment – it all needs to be there. And, since some of this training will be stressful (Figure 6.1), it makes sense to co-locate medical facilities to check the health of the future astronauts and certify them for spaceflight. This will be especially true in the early stages of the industry, because wealthy individuals, who can afford the flights, tend to be older and less healthy than average. There will also need to be emergency facilities in case of accidents.

After a hard day's training, our astronauts-to-be and their friends will want to kick back and relax, so hotels will need to be built near, or attached to, the spaceports. Staying with the relaxation theme, it will make sense to co-locate entertainment facilities, so family and friends can occupy themselves during the training. Perhaps an IMAX-type theater would be an attraction? Or a space theme park, with rides and space simulations? If these entertainment facilities are well designed, they could be a destination in themselves, even when there are no launches taking place. For example, the idea of a Space Camp/Academy is a great way to get kids involved and provide them with the opportunity to learn about the suborbital flight experience.

Table 6.1. Spaceport Features.

Class	Feature description
Local infrastructure	Runway
	Railhead
	Road access
	Hotels, restaurants, and shops
	Qualified local workforce
	Proximity to university
Site facilities	Pads for sounding rockets
	Pads for small, medium, and large sRLVs
	Horizontal take-off/landing capability
	Fuel handling/solid
	Fuel handling/liquid
	Fuel handing/hybrid
	Chemical analysis facilities
	Ordnance facilities
	Vehicle integration/checkout
	Payload processing – hazmats
	Processing – dynamic balance
	Spacecraft storage facilities
	Engineering/mission management offices
	Range radars, cameras
	Telemetry data retrieval
	Payload processing-vibration
	Engine test stands
	Materials testing facilities
	Hazmat training
	On-site research labs
	Broadband access
	Emergency response teams
	Downrange payload retrieval
Space training	Medical facilities
	Training facilities
	Simulators
	Space Academy
	Family facilities/residential
	Family facilities/entertainment
Financial/admin	Financial incentive/trade zones
	International facilities/customs
	Security for military users
	High-tech company incubators
	Simplified admin (safety, environment)

Public access to witness the launches will be needed – a place where members of the public can stroll in and out while launch preparation is taking place, and where they can watch the events unfold. And, in the event of any delays, there should be ample restaurant facilities and souvenir shops (Table 6.1).

SPACEPORT AMERICA

The first spaceport to be built from the ground up is Spaceport America (Figure 6.2).[1] Thanks to the fact it is home to Sir Richard Branson's Virgin Galactic, the world's first commercial spaceline, the US$209 million project has attracted worldwide attention. Designed, built, and operated by the New Mexico Spaceport Authority (NMSA), Spaceport America's operational infrastructure includes an airfield, launch pads, a terminal/hangar facility, emergency response capabilities, utilities, and roadways. The site will be capable of accommodating the activities of vertical and horizontal take-off vehicles, serving as a base for astronaut training, and providing a tourism experience for families and friends of those with a ticket to ride.

If you're interested in visiting, Spaceport America is located west of the US Army White Sands Missile Range in Sierra County, in New Mexico, or about 50 kilometers south-east of Truth or Consequences. The spaceport is easily accessible by county roads from Interstate-2 and it has been operational for a while; several flight tests have taken place since 2006. Overseeing the spaceport's development is the NMSA, the state agency that built the structure using funding provided by the state of New Mexico. Headquartered at the spaceport is Virgin Galactic, the anchor tenant.

6.2   Spaceport America. Courtesy: Spaceport America

---

[1]   www.spaceportamerica.com.

While Spaceport America may look like a futuristic airport, it's not the sort of place where you can land your Citation jet because it is designated to operate as a prior permission required (PPR) airport, which means there are no services for general or commercial aviation. Another downside is that there are no commercial airline flights to Spaceport America, so when you're planning your rocket ride, you will have to fly into either El Paso International Airport (ELP) or Albuquerque International Sunport (ABQ).

If you're searching for possibly one of the coolest locations on the planet to hold your next annual conference, Spaceport America can handle it.[2] NMSA's office is the building just west of Starbucks, and their office is on the second floor.

Although no rockets have taken off carrying fare-paying passengers, the spaceport is already a hub of activity thanks to residents such as Virgin Galactic and Armadillo Aerospace. Keeping them company is UP Aerospace, which was the first company to launch a commercial payload from the spaceport. UP Aerospace offers payload recovery and a range of tracking, telemetry, and avionics options. Another significant presence is Lockheed Martin, which, as one of the nation's top aerospace companies, needs no introduction. The aerospace behemoth has been using the spaceport as a base to test new launch and recovery technologies.

CARIBBEAN SPACEPORT

For those who would like to combine their trip of a lifetime with another destination vacation, there's probably no better place than the Netherlands Antilles island of Curaçao, home to the Caribbean Spaceport (CSP).[3] Originally conceived in 2005 in cooperation with various spaceflight and business professionals, the spaceport is now run by Spaceport Partners, who work closely with governmental, academic, and business institutions to research and assess the technological, legal, and economic feasibility of developing and operating the spaceport. We'll get to latest developments shortly, but first some history.

Although the idea for the spaceport was dreamt up in 2005, it took a while to put together the necessary feasibility studies, requirements analyses, business planning, and architectural design that involved the TUDelft, the University of Leiden, the Dutch Government, and DDOCK Design. In fact, the ball only really got rolling in August 2008, when the Caribbean Spaceport venture spent two weeks on Curaçao to present its plans to government officials, local business people, and the general public. The idea received a warm welcome, prompting CSP founder and director Joost Wouters to extend an invitation to Buzz Aldrin, a group of NASA astronauts, oceanographers, and business executives in the SeaSpace group for another presentation in October 2008. This presentation was

---

[2] Just contact NMSA's offices at 901 E. University Ave, Suite 965L, Las Cruces, NM 88001, USA, phone: 575-373-6110.

[3] www.caribbeanspaceport.com, Caribbean Spaceport, Sphinx Building, Baron G.A. Tindalplein, Suite #185, 1019 TW Amsterdam, The Netherlands, e-mail: info@CaribbeanSpaceport.com, phone: +31-(0)6 123-66-000 or +31 (0)6-506-07-110, fax: +31 (0)20-776-2775.

followed by a February 2009 visit by Sir Richard Branson, who showed great enthusiasm for the Caribbean Spaceport concept and requested to take the design brochure with him. In April 2009, Wouters presented a lecture about Commercial Spaceflight and Spaceports during the TUDelft VSV International Entrepreneurial Spaceflight Symposium "Ready to Launch" presentation, which included talks by Odyssey Moon CEO Bob Richards and XCOR COO Andrew Nelson. Two months later, Buzz Aldrin showed his support for CSP at the International Space Development Conference (ISDC) in Orlando, Florida. The rest of the year included meetings with Virgin Galactic CEO Will Whitehorn, who expressed Virgin's positive position, and meetings with Bigelow Aerospace, SpaceX, Masten Aerospace, XCOR, and NASA. Word was getting around, and the media were gradually picking up on the CSP venture.

Today, CSP has concluded its feasibility studies, requirements analyses, and business planning, and is in the process of discussing investment options and acquiring funding. It is also in contact with various operators and spacecraft developers concerning future operations from its spaceport. If all goes well, the spaceport will be open for business in 2014. When complete, CSP will offer all the facilities necessary for training suborbital passengers, a SpaceExpo, entertainment, bars and restaurants, and a shopping mall for friends and family accompanying the space tourist.

Compared to many other spaceport locations, CSP location offers a number of advantages, one of which is using the existing high-tech infrastructure of Hato International Airport of Curaçao. Hato's 3.5-kilometer runway is the longest in the Caribbean and is more than long enough to deal with launches for the suborbital spacecraft currently in development. Secondly, unlike many other remotely located spaceports, Curaçao offers an attractive setting with a fully developed tourism infrastructure.

SPACEPORT SWEDEN

Tropical weather and tequila not your style? No problem. Just take a flight to Stockholm and head north to Kiruna, home of the Spaceport at the Top of the World.[4] Given its location, the Swedish town of Kiruna may seem an unlikely place to build Europe's first commercial spaceport. Its 67.86° latitude means it is 150 kilometers above the Arctic Circle and close to 900 kilometers north of Stockholm. In addition to its extreme grid reference, Kiruna has a number of other disadvantages; it is home to the world's largest underground iron ore mine (that is expanding relentlessly), a vast expanse of forests, no sunlight for weeks at a time (from the first week of December until the second week of January, Kiruna has zero hours of sunlight per day), and temperatures that are great for polar bears, but not so good for tourists (the average high in July is just 7°C). None of these shortcomings stopped the Swedish government though. In 2007, it announced an "agreement of understanding" with Virgin Galactic to make Kiruna the company's first launch site outside the US. If all goes to plan, Sir Richard Branson could soon be flying his space tourists on suborbital jaunts straight through the aurora borealis (Figure 6.3).

---

[4] www.spaceportsweden.com, e-mail: info@spaceportsweden.com, Twitter: @SpaceportSweden, phone: +46 (0) 980 80 880, Mon–Fri 09:00–17:00hrs CET.

6.3 Aurora borealis. Courtesy: NASA

While Spaceport Sweden will be new to many tourists, Kiruna is not completely undiscovered. The town has been home to an array of aerospace activities since the Swedish government established a space research center there in 1964. The center – Esrange – includes a 5,600-square-kilometer range for launching sounding rockets (Figure 6.4). While launching the odd satellite offers a little excitement for the locals, it's the prospect of people flying into space from their snow-covered airfield that has Kiruna's residents' attention. With typical Virgin panache, the suborbital Kiruna flights are being promoted as an Arctic adventure complete with a stay in a hotel made of ice (Figure 6.5) and snowmobile rides through the wilderness.

I visited Kiruna many years ago during a backpacking trip around my native Scandinavia (I was born in Norway). Back then, except for iron ore, the town had few natural resources. This wasn't the place you came to take a skiing trip (no mountains, or hills for that matter) but if reindeer watching was your thing, you'd have had plenty to do. But that was more than 20 years ago. Today, thanks to Kiruna's talent for marketing, the idea of marrying tourism with space has the backing of just about everyone in the town, which isn't surprising because the residents know they can't just live on income from the mine; with increasing automation, the town is steadily shrinking. And, with a dwindling population, people realized they needed to have other businesses, so the town began investing in tourism, to attract visitors during Kiruna's short summer. The investment paid off for six months of the year; white-water rafting, fly-fishing, survival training, and canoeing were big draws. But for the other half of the year the town's hotels were empty. Then came the IceHotel (www.icehotel.com), a hotel with rooms built out of snow and ice (*snice*).

6.4 Rocket launch from Kiruna's Esrange. Courtesy: Esrange

6.5 Inside Kiruna's IceHotel. Courtesy: Wikimedia

Guests pay between US$200 and US$600 to spend a night bundled in sleeping bags on reindeer skins in a sub-zero room. Now, instead of empty hotel rooms in the dark months of December and January, this period is the town's high season.

Building a hotel out of *snice* and charging US$600 a night might have struck many to be too outrageous a business plan to succeed, but it did, which is probably why Kiruna is so supportive of Virgin Galactic's equally outrageous plans; like the IceHotel, suborbital spaceflight could mean another tourist boom in Kiruna. After all, let's not forget the sort of tourist we're talking about here. Those rich enough to splash out US$250,000 on a ticket will most likely want their nearest and dearest along to share the experience, which all adds up to lots of hotel rooms. And these people all need activities to occupy themselves while their soon-to-be astronaut spends three days training for their spaceflight. Space summer camps. Centrifuge runs. Reindeer watching. It's all part of Spaceport Sweden's marketing strategy.

The IceHotel's marketing department, recognizing a good business opportunity when they see it, is already working with Spaceport Sweden to come up with a plan to fuss over Virgin's customers, from accommodations and entertainment to working with Virgin's medical staff to produce just the right preflight menu for their four-star restaurant. So that's the tourism infrastructure taken care of. But what about the business of actually sending rockets into space? Well, thanks to Esrange, most of that infrastructure is already in place. Esrange stages between five and 10 launches per year, and is the European Space Agency's primary site for launching research rockets. It's also been in the business of suborbital flights since 1966, albeit unmanned (the range's MAXUS and MASER rockets

6.6   Mojave Air and Space Port. Courtesy: Mojave Air and Space Port

offer up to 13 minutes of microgravity at the peak of their suborbital flights). Once manned suborbital flights begin from above the Arctic Circle, they'll take off from Kiruna Airport, which has just four departures a day (the lack of air traffic is another of Kiruna's competitive advantages, since free slot time means there is plenty of time for launching spacecraft).

While it would appear Spaceport Sweden is almost open for business, there are still some issues that must be addressed, including regulatory approval from the US[5] which

---

[5]   An example of this regulatory approval is placing spacecraft on the United States Munitions List (USML). For years, the US commercial space industry fought to remove export restrictions placed on it during the 1990s. In 2013, change was on the horizon, but it was a case of one step forward, two steps back because man-rated suborbital spacecraft were added to the list! This is a problem because any item on the USML requires an export license from the US State Department. Even worse, putting suborbital spacecraft on the USML places them under the restrictive umbrella of the International Traffic in Arms Regulations (ITAR). Why is this being done? The 1990s restrictions were intended to block the flow of space technologies to nations such as China and to maintain US space competitiveness. The upshot of this was that the restrictions harmed rather than strengthened the US commercial satellite industry; US satellite makers were denied access to foreign markets and lower-cost launchers for their products. The result? A significant US share of the global commercial satellite market was lost to China! And you wonder why politicians are unpopular! But, in December 2012, after years of lobbying by the US satellite industry, a provision in the 2013 defense authorization Bill passed by the US Congress struck out the 1990s language that placed satellites and related items on the USML.

may have concerns about exporting technology with potential military applications and Swedish authorities. But, assuming flights take off in New Mexico, then it shouldn't be too long before we see spacecraft taking off from the Arctic.

MOJAVE SPACEPORT

Here's a trivia question for you – what do these films have in common: *Die Hard 2*, *Executive Decision*, *Flight Plan*, *The Stand*, *Thirteen Days*, *Tuskegee Airmen*, and *Waterworld*? Answer: parts of them were filmed at KMHV, also known as Mojave Airport (Figure 6.6),[6] a sort of ground zero for commercial suborbital spaceflight. Housed in a collection of dusty hangars and sheds are Scaled Composites, XCOR Aerospace, Masten, and The Spaceship Company. In short, Mojave Space Port is a Mecca for aviation geeks; it offers tours to the public on weekdays, so if you ever find yourself in the high desert with some extra time, pay a visit. Time your visit right and you just might see SpaceShip Two (SS2) making a test flight. Once you've got your fix for exotic spacecraft, you can wander down to the Mojave Airport Park and walk past a line of F-4 Phantoms, a Lockheed L-1011, or even a 1950s era Grumman Albatross amphibious airplane. Also at KMHV is the National Test Pilot School (NTPS) and on the far side of the runway are rows of 747s and 737s, parked during slow times or waiting to be cannibalized for spare parts.

Located a two-hour drive north of Los Angeles, Mojave Air and Space Port has become one of the most iconic locations in the commercial suborbital industry. Home to 14 space companies, this vast expanse of flat, scrubby desert has witnessed *thousands* of rocket tests, although the only vehicle that has flown into space is SpaceShipOne (SS1), a ship that is the model for Virgin Galactic's space tourism venture. Spoken of by some as the Silicon Valley for commercial spaceflight, this is a place where test pilots still push the envelope of aerospace technology. And if that sounds a little like *The Right Stuff*, it's worth bearing in mind that Chuck Yeager first broke the sound barrier in the X-1 rocket plane at nearby Edwards Air Force Base in 1947. And the first orbital Space Shuttle flight landed there as well. With wide-open stretches of desert, the long runway, and few neighbors to complain about the noise, it's easy to see why companies have been drawn to Mojave, although with its dusty warren of wind-worn hangars and sheds, the place doesn't exactly scream spaceport: stand in the center of Mojave and look around and all you'll see is a desert landscape dotted with sage brush and Joshua trees.

While the Mojave Airport and Space Port has led the way in the spaceport industry since it was the site of the first privately funded spaceflight in 2004, the new kid on the block – Spaceport America – has established the template for future commercial space-ports. But that's not to say Spaceport America doesn't have competition. On the other side of the world, there are plans for a Spaceport Singapore, funded by financial assistance from the United Arab Emirates. Nor does Spaceport America's glamour discount the usefulness of other commercial spaceports in Oklahoma, Alaska, and Florida; these

---

[6]   Mojave Air and Space Port, 1434 Flightline, Mojave, CA 93501, USA, e-mail: info@mojaveairport. com, phone: (661) 824 2433.

spaceports were converted from old federal launch sites or airports and were not created solely for the suborbital space industry. Instead, they fill the role of launch service providers that send commercial and research satellites into orbit. How successful will these spaceports be? It's difficult to say, but the leading contenders in the US seem to be Mojave and Spaceport America. What is clear is that a spaceport must cater to the needs of commercial astronauts, space tourists, and their families and friends. It is also clear that the suborbital space industry won't really take off without a parallel effort that builds the spaceport architectures to support the business terrestrially.

7

Suborbital Science

"You spark this industry with tourists, but I predict in the next decade the research market is going to be bigger than the tourist market."

Dr. Alan Stern, Southwest Research Institution

When people think about manned suborbital flight, they usually conjure up images of tourists suited up in those very cool Virgin Galactic suits flying around the interior of SpaceShipTwo (SS2). It's a popular image that has permeated the public consciousness thanks to Virgin's publicity campaign touting tickets to space. But while space tourism will surely be big business for Virgin Galactic, science, perhaps even more than tourism, could become an even bigger income earner. Virgin is selling tickets for US$250,000, which is peanuts compared to the hundreds of thousands of dollars that some scientific space projects can cost. And, if you don't have US$250,000 lying around, how about flying on the Lynx, for just US$95,000?

The attraction of commercial suborbital science missions isn't just low-cost; there is also the appeal of increased instrument flexibility since vehicles can support unmanned payloads and human-tended experiments, with different vehicle types providing a variety of mission profiles. The new fleet of suborbital vehicles also have unique fly-on-demand, rapid-turnaround, and human-in-the-loop capabilities that will enable new types of previously impossible research. One of those eyeing the potential of suborbital science is NASA, which formed its Flight Opportunities Program (https://flightopportunities.nasa.gov/) to seek proposals for suborbital technology payloads and spacecraft capability enhancements that could help revolutionize future missions. Here's what a couple of students had to say about the program:

"NASA Flight Opportunities/Parabolic has been a great opportunity for us to gain access to the unique microgravity conditions that will allow us to get the data needed to develop more efficient heat exchangers."

Jungho Kim, University of Maryland

E. Seedhouse, *Suborbital: Industry at the Edge of Space*, Springer Praxis Books, DOI 10.1007/978-3-319-03485-0_7, © Springer International Publishing Switzerland 2014

7.1 Alan Stern after his ride in the Starfighter. Courtesy: Southwest Research Institute

"The microgravity test environment allowed us to understand more fully the behavior of several deployment strategies in a way that is not possible in the laboratory environment."

Sigrid Close, Stanford University

Other organizations promoting suborbital science include the Commercial Spaceflight Federation (CSF), which has organized science workshops in conjunction with NASA Ames and the Universities Space Research Association (USRA), to highlight the research and education potential of suborbital launch vehicles. The CSF has also created an advisory panel, the Suborbital Applications Researchers Group (SARG), composed of experienced scientists, researchers, and educators dedicated to promoting the usefulness of commercial suborbital research flights. SARG is chaired by Dr. Alan Stern (Figure 7.1) of the Southwest Research Institute (SwRI), a space scientist who previously served as head of the Science Mission Directorate at NASA Headquarters. Alan hit the headlines at the 2012 Next Generation Suborbital Researchers Conference (NSRC) when he announced that SwRI had budgeted US$1.3 million for a four-year suborbital science program, a portion of which will be used to book passenger seats on spacecraft (SwRI bought eight seats with options for nine more on suborbital flights, split between Virgin Galactic's SS2 and XCOR's Lynx) for Alan and his SwRI colleagues, Dan Durda and Cathy Olkin. If all goes as planned, the three researchers will be flying into space as fully fledged astronauts by mid-2015.

POTENTIAL SUBORBITAL SCIENCE CAPABILITIES

- High flight rates with rapid-turnaround and fly-on-demand capability permit intensive support for unmanned payloads and astronaut-tended research
- Lower cost than sounding rockets and other methods
- Potential research fields include Earth science, heliophysics, planetary sciences, astronomy, microgravity physical sciences, life sciences, aeromedical
- Affordable platform for flight qualification and raising hardware technology readiness levels (TRLs)

It's not only research organizations like SwRI that are getting in on the suborbital science market. In 2009, Aabar Investments, the Middle East investment fund that bet big on Mercedes-Benz (it bought 9.1% of the German auto maker in March 2009) paid about US$280 million to buy nearly a third of Virgin Galactic. Aabar Investments' buy-in gave Branson's space tourism venture a big financial kick-start at a time when many funding sources had dried up because of the global recession. It also gave the wealthy Persian Gulf sheikdom of Abu Dhabi a chance to build its own spaceflight industry as it broadened its economy beyond the oil sector. In exchange for its 32% stake in Virgin Galactic's holding company, the state-controlled fund will acquire "exclusive regional rights" to eventually launch Virgin Galactic scientific research spaceflights from the United Arab Emirates (UAE) capital.

More recently, Virgin Galactic and Aabar Investments were selected by NASA to provide flight opportunities for researchers to fly technology payloads into space. This arrangement not only marked the first time NASA had contracted with a commercial partner to provide spaceflights on a suborbital spacecraft, it also represented an endorsement of the value of regular commercial space access for a range of science applications. Although generally regarded as a space tourism company, Virgin Galactic has recognized that providing access to space to researchers and their experiments is an important future mission segment and a significant business opportunity, which is why SS2 can be configured for tourists or scientists (Figure 7.2). Virgin Galactic has also assembled an expert team of partners to provide payload integration and flight services for the impending science missions, including Southwest Research Institute of Texas, NanoRacks, LLC of Texas, and Spaceflight Services of Washington. These partner organizations bring extensive experience flying scientific experiments on suborbital and orbital rockets, as well as the International Space Station (ISS).

PORTFOLIO OF GAME-CHANGING MISSIONS

Attendees at the annual NSRC will lead you to believe there is a veritable bounty of science missions waiting to be tapped by the commercial suborbital spaceflight industry – everything from heliophysics to high-altitude meteorology and biophysics to bioscience. One very popular suborbital science topic is atmospheric research. In fact, one of the first manned suborbital science missions to be flown on the Lynx will be the PoSSUM (Polar Suborbital Science in the Upper Mesosphere) mission investigating the small-scale dynamics of noctilucent clouds (Figure 7.3). We'll return to PoSSUM shortly.

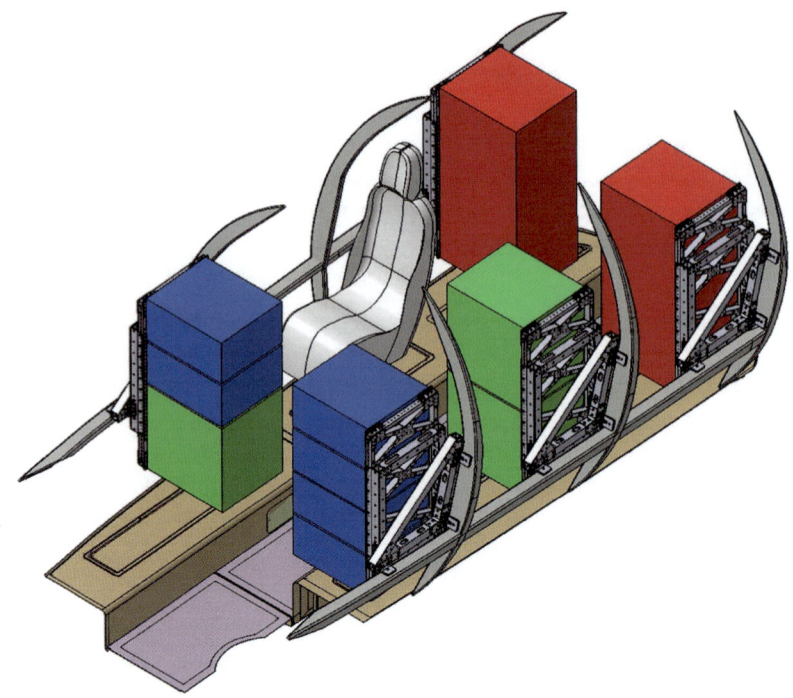

7.2   Virgin Galactic payload configuration. Courtesy: Virgin Galactic

7.3   Noctilucent clouds. Courtesy: NASA

7.4   Terrestrial gamma-ray flashes. Courtesy: NASA

Another research topic requiring suborbital mid-altitude experimentation is the investigation of Terrestrial Gamma-ray Flashes (TGFs). Discovered in 1994, TGFs (Figure 7.4) are bursts of gamma rays in Earth's atmosphere thought to be caused by electric fields produced above thunderstorms. Recent observations show that approximately 50 TGFs occur each day – a number that may be much higher due to the possibility of flashes in the form of narrow beams that would be difficult to detect, or the possibility that a large number of TGFs may be generated at altitudes too low for the gamma rays to escape the atmosphere. To investigate this possibility, suborbital access is ideal, since it permits frequent measurements, tailored launch windows, and flexible flight plans that will allow scientists to hand-pick the storms they want to study.

Planetary scientists are also eyeing the possibilities offered by suborbital flight – among them, SwRI's Alan Stern and Dan Durda, who are interested in dust aggregation and condensation in early planetesimals. Alan and Dan reckon the aggregation and condensation process can be studied in suborbital flight because inter-particle gravitational forces are negligible and collision speeds are low. A related planetary science mission is in the field of space-based observation, which is why the Planetary Science Institute (PSI) and XCOR Aerospace have signed a memorandum of understanding that lays the groundwork for flying the human-operated Atsa Suborbital Observatory aboard XCOR's Lynx spacecraft.

The Atsa (which means "eagle" in the Navajo language) project will use the Lynx equipped with a specially designed telescope to provide low-cost space-based observations of Solar System objects near the Sun that are difficult to study from orbital observatories and ground-based telescopes. To ensure optimum data acquisition, the Lynx spacecraft will fly a modified flight trajectory and will be capable of precision pointing, allowing the Atsa system to acquire the desired target and make the planned observations.

In a related field, heliophysicists hope to use suborbital vehicles as a platform to study high-frequency waves and Doppler shift in the solar chromospheres. A similar project flew on the Shuttle, but with the retirement of the Orbiter, human-in-the-loop opportunities no longer exist. Now, with routine suborbital access on the horizon, heliophysicists plan on flying special telescopes and solar pointing systems to demonstrate UV imaging from the platform. Once the UV imaging has been proven, the heliophysicists plan to raise the technology readiness level of their equipment and fly a new set of detectors.

Perhaps the field with the most potential, given the number of tourists that will be flown over the next few years, is human physiology. Don't forget that our experience in manned suborbital flight extends to half a dozen manned flights and the odd animal flight. With so many tourists and scientists flying, it will be important to characterize neurovestibular responses (the "stand test"), oculomotor control and perceptual disturbances, motion sickness onset, and fluid shifts. To characterize these responses, passengers will be instrumented for heart rate, electrocardiogram, arterial blood pressure, oxygen saturation, blood volume, and electroencephalogram activity.

The stand test

Even during a suborbital flight, the body will undergo changes that may make the transition back to Earth's gravity a little challenging for some. For example, some crewmembers may find it difficult standing upright due to orthostatic intolerance. Such a simple task may prove problematic because it is in this position that gravity is exerting most of its influence on the cardiovascular system. Think of an upright human body as a tall column of water. As gravity is pulling down on that column of water, each level, or depth, of water is influenced. In any body of water, the pressure at the surface of the water is equal to atmospheric pressure, but the pressure rises by 1 mmHg for each 13.6-millimeter distance below the surface. This pressure results from the weight of the water above it and is called *hydrostatic pressure*. Hydrostatic pressure also occurs in our vascular system because of the weight of the blood in the vessels. Hydrostatic pressure exists on Earth because of gravity. When a passenger spends time in suborbital space, their cardiovascular system will try adjusting to functioning without gravity. For those with a poor cardiovascular system, the return to Earth's gravity may prove challenging. To test a passenger's orthostatic intolerance, life scientists have suggested administering the "stand test" pre and post flight (pre-flight measurements will serve as baseline control measurements from which to compare the post-flight data). The measurements can give scientists an indication of what may be happening to cause orthostatic intolerance and help them develop countermeasures. The stand test consists of a 29-minute supine period during which the passengers are instrumented to measure heart rate and blood pressure cuff. Next, the passengers are asked to stand for an 10-minute period, during which measurements are taken again.

In addition to characterizing all these responses, scientists will want to start building a database comparing passenger profiles with respect to fitness levels, smoking, stress, body mass index, high blood cholesterol, and inactivity. They will also want to compare responses based on age, gender, race, flight experience, and various medical conditions, and compare these factors against vehicle trajectories and the cumulative effects of repeated suborbital exposures. Another interesting study will be assessing transitions between various G-loading "push–pull" effects from single and repeated exposures. In short, there will be plenty of projects to keep scientists busy, but how does one go about flying one of these projects? To answer that question, let's tell the story of someone who has actually gone through the process. Meet Jason Reimuller.

ANATOMY OF A SCIENCE MISSION

Dr. Jason Reimuller (Figure 7.5) is a Research Scientist with the Space Science Institute and President of Integrated Spaceflight Services (www.integratedspaceflight.com). He spends much of his time developing spacecraft egress training modules, training

7.5 Jason Reimuller, Principal Investigator of the PoSSUM flight. Jason holds a Ph.D. in Aerospace Engineering Sciences from the University of Colorado in Boulder. He also holds an M.S. degree in Physics, Aviation Systems, and a B.S. degree in Aerospace Engineering. Courtesy: Jason Reimuller

7.6   Polar mesospheric clouds. Courtesy: NASA

simulators, and analog space training capability using aerobatic aircraft as part of an integrated space research and training facility. Jason also works as a commercial research pilot and flight test engineer with GATS, Inc., and, when he's not doing that, he serves as Chief Operating Officer for Astronauts for Hire (A4H).

Jason has plenty of experience working in the industry, having served for six years as a system engineer and project manager for NASA's Constellation Program, leading studies on launch aborts, launch-commit criteria, landing conditions, post-landing and emergency crew egress trades, and propulsion options. He also led a NASA- and NSF-funded flight research campaign to study noctilucent cloud time evolution, structure, and dynamics in Northern Canada as lead investigator and pilot-in-command, and it was in the field of atmospheric science that Jason had his idea for a suborbital mission.

Dubbed PoSSUM (Appendix IV), Jason's flight will study polar mesospheric clouds, or PMCs. PMCs (Figure 7.6) are the highest clouds in Earth's atmosphere, and are observed slightly below the mesopause in the polar summertime. The clouds are of particular interest because they are sensitive to global climate change. Thanks to satellite and ground-based observations over the past four decades, scientists now realize these clouds are sensitive indicators for what is going on in the atmosphere at higher altitudes as small changes in the atmospheric environment can lead to large changes in the properties of these clouds.

Jason's idea was to use the capability of manned suborbital vehicles to answer certain questions about PMCs. For example, what are the small-scale dynamics of PMCs and what does this tell us about the energy and momentum deposition from the lower atmosphere? And what is the seasonal variability of PMCs, mesospheric dynamics, and temperatures?

7.7   PoSSUM patch. Courtesy: Jason Reimuller

In addition to trying to provide data that may help answer these questions, PoSSUM (Figure 7.7) will also validate a method to conduct repeatable, low-cost, in-situ PMC observations. A secondary objective of the flight is to conduct follow-up flights to mature the instrument suite used on PoSSUM into an integrated, modular laboratory that will help the ground observation, glaciology, forestry, and aeronomy research communities.

The PoSSUM mission will be flown on the XCOR Lynx Mark I from a high-latitude spaceport. When strong cloud formations are observed from the ground, the Lynx will be launched to an altitude that transitions the PMC layer. The clouds will be directly illuminated from the Sun and the Lynx's attitude will be oriented north to the estimated region of highest cloud density. Instrumentation will include video and still-frame cameras, an infrared camera, a mesospheric temperatures experiment, and a mesospheric winds experiment. Making sure everything works will be Jason, who will be sitting in the right seat.

So, let's assume you've been successful winning a flight for your experiment. What sort of timeline can you expect to follow before actually strapping into the Lynx or SS2? A generic timeline is outlined in Table 7.1, and begins with the appointing of a Principal Investigator (PI) who will embark upon a research design for the flight, which could take up to six months or longer. Some time during the research design period, the PI (this could be you) will receive an experiment form from the operator, requesting that the PI provide them with the particulars of all the scientists involved in the experiment, experiment objectives, experiment description, and a technical description of the experiment set-up. This will require the PI to provide a description of each system, an explanation of each experiment rack (including designation, function, mass, dimensions, etc.), schematics of the experiment, details of electrical circuits, a list of all products, photographs of the experiment set-up, and, finally, the team's approach for designing, building, and testing the experiment. Other items on the experiment form may include details such as electrical

Table 7.1. Timeline for a Suborbital Science Flight.

	L–36	L–30	L–24	L–18	L–12	L–6	L–3	Launch
Submit proposal	▲							
Proposal review	◄———►							
Proposal acceptance			▲					
Research design			◄————————————►					
PI receives experiment form				▲				
PI confirms participation					▲			
Medical exam				◄——►				
Operator receives experiment form						▲		
PI and scientists visit operator							▲	
Execute changes to research							◄——►	
Design frozen								▲

Table 7.1. Timeline for a Suborbital Science Flight.

Medical docs submitted to operator								▲	
Test of payload								▲	
Safety review meeting									
Changes if necessary									
Liability form submitted to operator									
Preflight training								⬌	

consumption, a mechanical resistance analysis, in-flight procedures (a list of major tasks the scientists will perform during the flight), certification for use of human subjects, a liability waiver, an experiment hazard evaluation, and a hazard list. It's a lot of paperwork, but it doesn't stop with the experiment form because, two months before the flight, the equipment data package form arrives. This requires more specific information including a structural load analysis, proof of mechanical resistance of each structure, detailed test procedures, the particulars of the data acquisition system, and test operation limits and restriction.

Once that round of paperwork is out of the way, the team can look forward to the Safety Visit. The Safety Visit is the final review before the flight and allows the operator to inspect test equipment, check any modifications to the payload, and approve or disapprove the test equipment. Around this time is when the team has to submit their medical certificates and insurance (Appendix IV) to the operator. After the Safety Visit, the team head back home and work on any modifications to the payload. Then, a week before the flight, they pack their bags and travel to the spaceport to begin their preflight preparations. One of the first items on the schedule will be more paperwork; the operator will verify all

Table 7.2. Suggested Events Leading Up to Suborbital Science Launch.

Time	Event
L–7 days	Scientists and team members arrive at spaceport. Review by Principal Investigator
L–6 days	Commence experiment preparation at Spaceport science staging facility
L–5 days	Configure experiment into spacecraft
L–4 days	Commence preflight training Day #1 of 4. AM: Academic instruction on high-altitude indoctrination and high acceleration. PM: Slow and rapid decompression in high-altitude chamber followed by emergency egress procedure training
L–3 days	Preflight training day #2 of 4. AM: Flight G profile in centrifuge followed by task acquisition exercise review. PM: Zero-G preflight familiarization followed by pressure suit acquaintance and testing (donning and doffing)
L–2 days	Preflight training day #3 of 4. AM: Survival brief
L–1 day	Preflight training day #4 of 4. AM: PM: Flight safety briefing followed by presentation of each experiment by respective PIs
L–1 hour	Scientists meet at Spaceport for final preflight debrief. Optional anti-emetic medication given. Scientists conduct final check of payload
L–30 minutes	Scientists and crew board spacecraft. Experiments are switched off
L–20 minutes	Hatches are closed. Passengers are requested to be seated and spacecraft begins taxiing. Spacecraft electrical panel switched off
L–10 minutes	Spacecraft electrical panel switched on
Launch	Spacecraft takes off
L+15 minutes	Passengers leave their seats and switch on experiments
L+20 minutes	Passengers switch off experiments and adopt landing configuration. Electrical panel switched off
L+35 minutes	Landing and debrief
L+4 hours	Modifications and preparation of experiments for following day

passengers have their medical certificates in order that liability and waiver forms, if required, are signed and all necessary modifications have been implemented. The operator will give the science team an overview of the week's activities and training. Once all the paperwork is out of the way, the scientists will begin positioning their payload in the spacecraft, assisted by the operator's checkout team who will help the scientists with attachment interfaces and electrical input and output requirements. The loading, bolting, and electrical connecting will take about two days, after which the team will commence preflight training. A suggested timeline is provided in Table 7.2.

MAKING THE MOST OF THOSE FOUR MINUTES

In case you manage to have your science experiment accepted, what can you do to make the most of those precious four minutes of weightlessness? Well, first of all, you would do well to apply the principles of the three "P"s – prior preparation and planning – but we'll get to that shortly, because all the preparation in the world will be for nothing if you can't avoid the dreaded space motion sickness/space adaptation syndrome (SAS). SAS is a subject space operators are loathe to mention because the last thing you want to be is sick as the proverbial dog during your US$250,000 suborbital flight. But, sure as eggs are eggs, SAS will be a problem given the provocative nature of suborbital flights (the boost acceleration particularly). One way to minimize the chances of green faces is to take anti-motion-sickness medication, but be wary of side effects. For example, promethazine has been used to treat SAS during Shuttle missions, but the side effects associated with this drug include dizziness, drowsiness, and impaired psychomotor performance, which obviously impact crew performance.

Eyes Versus Ears: A Primer on Motion Sickness

The currently accepted theory of SAS is that sensory conflict is to blame. Basically, information from your visual and vestibular systems is processed by the brain to match it. Your vestibular system (inner ear) is tuned to a 1-G environment and, when you move around, changes in your vestibular system match what you're seeing. But when you get on board an aircraft, your inner ear signals that you're moving, but your eye says you're sitting still because your body is not moving in relation to the seat you're sitting in. This causes sensory conflict, which often results in emesis, or vomiting. The good news is that about 30% of the population is naturally immune to motion sickness … in most conditions. Why some people are susceptible and others are not is a question scientists have been trying to answer for decades. The truth is, they just don't know. Some people who are sick as a dog in motion environments on Earth do fine in space, whereas those who you would expect to do well in space, such as aerobatic pilots, for example, get SAS.

The latest research has experimented with combinations of more heavy-hitting drugs, and the latest anti-SAS concoction is a combination of oral scopolamine, to suppress vomiting, and dextroamphetamine, to counteract scopolamine's potential to induce drowsiness. Tests on passengers in the "Vomit Comet" – a DC-9 – found that the combination reduced the incidence of motion sickness from 70% to about 12%. If you're still worried about suffering a bout of SAS, you can also try desensitization training, which uses exposure to motion in an artificial environment (spinning chairs) and biofeedback, in which subjects learn to control their own breathing, heart rate, and other physical responses, to help deal with motion sickness. The bottom line is there is no magic bullet that will guarantee you will not get sick, so the best advice is to take the medication, try to maintain a "1-G orientation" and avoid abrupt head rotations.

Let's imagine you're one of the lucky ones and you're about to enter the microgravity phase of your flight and you're feeling great. How are you going to make the most of those four very expensive minutes (about US$520 *per second* based on the latest price of a flight on SS2!)? Remember, you'll have one shot and one shot only to gather data and perform whatever science experiment you've been tasked with. No doubt you'll have practiced your routine dozens if not hundreds of times, so you don't have to worry about the sequence of events. The big killer here is the unknown, especially the sudden transition from boost phase to microgravity coast, which will be distracting no matter how much you anticipate it.

This transition will be compounded by all the activity going on in the cabin as other scientists start to prep for their experiments and then there's that jaw-dropping view through the windows.

Experienced astronauts often speak of rookie flyers having all the grace of the proverbial bull in a china shop, which inevitably leads to damage and a disruptive workflow. So, rookie astronauts are taught that "slow equals fast" – a mantra that will definitely apply within the confines of a suborbital cabin. Another important mantra to remember is that the only certainty about an extravehicular activity (EVA) is uncertainty (a quote attributed to the legendary Story Musgrave). In other words, despite all those tasks on your checklist, you have to plan for things to go wrong. You can minimize off-nominal events by extensive pre-visualization and thinking through every – EVERY – detail, which includes nominal and off-nominal events. You can also help yourself by knowing your environment intimately and keeping track of everything, using either Velcro, lanyards, duct tape, or all three! Another critical skill will be the ability to maintain your situational awareness, because your fellow space scientists won't appreciate getting kicked in the head as they try to prepare their payload racks. You'll also need to keep a firm grip on your "space brain" as your mind becomes saturated with visual and task overload; don't forget, you'll most likely be required to perform several tasks within what seems to be an impossibly small amount of time, so it will be critical you use tethers (Velcro works as well) to keep track of your gear. You will also want to use a detailed plan of action, placards, and cuff checklists to make sure you do everything you need to do in the sequence in which it needs to be done. And don't forget to plan for contingency because even the simplest devices fail and you have to know how to repair them quickly. It's all part of your operational planning; don't leave anything to chance. Anything. The next step in your training is flight preparation, which should include training in as high a fidelity environment as possible – ideally a spacecraft mock-up. Failing that, you can utilize 1-G bench-top payload training and pre-visualization of the full sequence of mission operations. If you need help, there are companies out there that can help you, such as Suborbital Training (see Chapter 2).

LAUNCH DAY

Launch day begins with breakfast served in the spaceport restaurant (Table 7.3). While some may have an image of spacefarers enjoying a breakfast of steak, eggs, and coffee, the reality will be somewhat different because this new breed of commercial astronauts will probably be too damn nervous to eat anything more than half a slice of toast! However, in the early stages of commercial spaceflight, breakfast will be a mandatory photo opportunity for the media people, so the soon-to-be astronauts will probably fake a carefree smile and pretend to eat. After the meal, the crew performs one final prelaunch briefing before reviewing the launch countdown status and weather forecast. Next, the crew visits the flight surgeon for one final medical examination before visiting the bathroom for a final attempt to void their bladders before suiting up.

Unlike NASA astronauts, whose first challenge wasn't dealing with the acceleration stress following launch or adjusting to weightlessness but donning the rather cumbersome launch and entry suit, our commercial astronauts only have to contend with slipping on a

Table 7.3. Launch Countdown Milestones.

Time		
T–HH	T–MM	Event
02	00	Crew eats breakfast and conducts media interviews
01	30	Crew suit-up begins assisted by suit technicians
01	00	Mission management team meeting
00	50	Avionics checkout and pilot walk-around of vehicle
00	45	Crew weather briefing
00	40	Crew ingress vehicle and commence preflight checks
00	35	Flight crew equipment stow
00	35	Communication activation
00	30	Crew communication checks
00	30	Science team ingress vehicle
00	25	Debris inspection
00	20	Ascent switch list
00	15	Pad clear of non-essential personnel
00	10	Final crew weather briefing
00	05	Hatch closure. Vehicle pressurized
00	05	Area clear to launch
00	04	Launch Director launch status verification
00	02	Crew closes and locks their visors
00	00	Launch

lightweight pressure suit topped off with a helmet; a couple of suit technicians will check the integrity of the suit and the communication system. While the crew suits up, the vehicle is inspected in the same way as a regular aircraft. Using cameras and sensing devices, the flight team methodically checks every surface of the vehicle, looking for abnormal temperature readings or anything else that might indicate a problem. Once they finish their inspection, they report to the Launch Director.

Once suited up, the team wanders over to the vehicle and finds their seats. No close-out crew for this group of spacefarers – just a simple tug of the straps and hooking up to the communication system. The hatch is closed and a moment later the crew's ears pop as the capsule is pressurized. Meanwhile, back in Mission Control, the Flight Director thumbs through the countdown manual; it documents not only the launch procedures, but also the details of processing the spacecraft, the set of countdown instructions that sequentially configure the vehicle for launch, the instructions to be followed in the event of a scrubbed launch, and pre-planned contingency procedures and emergency instructions.

Once the launch team reports to the Flight Director that there are no constraints to launch, the Launch Director gives his permission to proceed with the countdown. Meanwhile, inside the spacecraft, the crew performs radio checks, and listens to the sequence of countdown milestones. As the clock shows the T–10-minute mark, the astronauts check their straps one more time, listen to the endless series of acronyms being checked off by the flight controllers, and watch their pilot follow the checklist. In just a few minutes, the vehicle will lift off from the pad in a thunderous cloud of fire and smoke (we're imagining it's a Blue Origin flight). Two minutes before launch, the Mission Control Console reminds the crew to close their visors – a command confirming launch is imminent. The crew oblige, snapping the bail into place to seal the faceplate. With adrenaline pumping and their heart rate close to the red-line in anticipation, the crew tries to slow their breathing and control the attack of butterflies. For a moment, the thought that these could be the final moments of their lives enters their head, but their attention is quickly drawn to the expectancy of the vehicle lighting up. After years of imagining riding a rocket into space, the rookie astronauts are finally going to get their chance at wish fulfillment. Bodies tingling with anticipation, the rookies give each other a nervous thumbs-up.

At one minute to launch, the flight displays continue scrolling data as the astronauts think about their families and all the training and preparation that brought them to this pivotal moment in their lives. The astronaut's families are a short distance away, standing on the roof of the spaceport, and are probably more scared than the astronauts. Mission Control announces 10 seconds to launch. At 5 seconds to go, the attitude indicators on the flight display scroll to the correct launch attitude as the vehicle processes its final update.

Finally, the countdown reaches the endpoint. Ignition commands from the vehicle's flight computers commit the vehicle to flight as electrical energy is routed to the motor igniter. The igniter initiates the burn of the rocket motor. A rumble shakes the vehicle as the engines spew fire and roar to full power. On the flight displays, the computers scroll rapidly through engine checks and mark the mission time: zero. The clock is running! Inside the vehicle, the astronauts are pushed back into their seats by a force more than one and a half times the force of gravity. The rocket engine is already consuming propellant voraciously as the crew is pushed further back into their seats as the gravitational forces ramp up. On the roof of the spaceport, the families watch awestruck as the vehicle blazes a trail of flame against the blue sky.

Less than 30 seconds after launch, shock waves form on the nose and the vehicle begins to vibrate. Meanwhile, the vehicle's control system continuously monitors trajectory and issues guidance commands. A few seconds later, the commander announces Mach 1 and the crew listen to the rushing sound of supersonic air being displaced by the vehicle. The commander gives the crew the signal they can shut off the oxygen and open their visors. The vehicle is now more than two minutes into flight and flying at an altitude of more than 60 kilometers. A few more seconds pass and the commander announces an altitude of 100 kilometers. The rookies let out a cheer, since the altitude means they are now officially astronauts and have collectively achieved their life goal. Then they go to work conducting their assigned mission science tasks.

8

Payloads

One headache suborbital scientist astronauts face is how to install their payloads in the vehicles, because there are all sorts of payloads and an assortment of vehicle designs. Take XCOR. Its Lynx spacecraft offers a variety of payload accommodations, one of which takes the place of the passenger's seat and would be large enough to accommodate shuttle middeck lockers (Figure 8.1). As you can see in the graphic, there's also a smaller area inside the cabin behind the pilot's seat. In addition to the internal payload space, the Lynx also offers a "ski rack" option, which is an external payload carrier that could be used for experiments requiring exposure to space.

Blue Origin is developing its own standard for payloads to make it easier for potential research customers. The company has developed what it calls the cabin payload bay, a standard module designed to accommodate a range of payloads, with a set of power and data interfaces. Blue Origin's approach towards payload integration is a plug-and-play capability, because the company prefers that researchers dedicate their time to their experiment rather than to a logistically difficult integration process. Then there is Virgin Galactic, which has signed a deal with NanoRacks, a company that has developed modular lab equipment that standardizes research payloads for the International Space Station (ISS). The deal provides that NanoRacks will develop standardized laboratory racks for SpaceShip Two (SS2). By developing standardized research payloads, Virgin will make it easier for scientific researchers to utilize the spaceship for microgravity experiments.

FLYING A PAYLOAD WITH VIRGIN GALACTIC

Virgin reckons each SS2 flight will be able to hold up to 590 kilograms of scientific payload and the rack configuration will include window access for the collection of atmospheric and spectrographic data. The racks will also be designed to allow non-standardized scientific payloads. To give you an idea of what is involved, Virgin Galactic explain the payload checklist in the Payload User Guide (PUG; Figure 8.2).

E. Seedhouse, *Suborbital: Industry at the Edge of Space*, Springer Praxis Books,
DOI 10.1007/978-3-319-03485-0_8, © Springer International Publishing Switzerland 2014

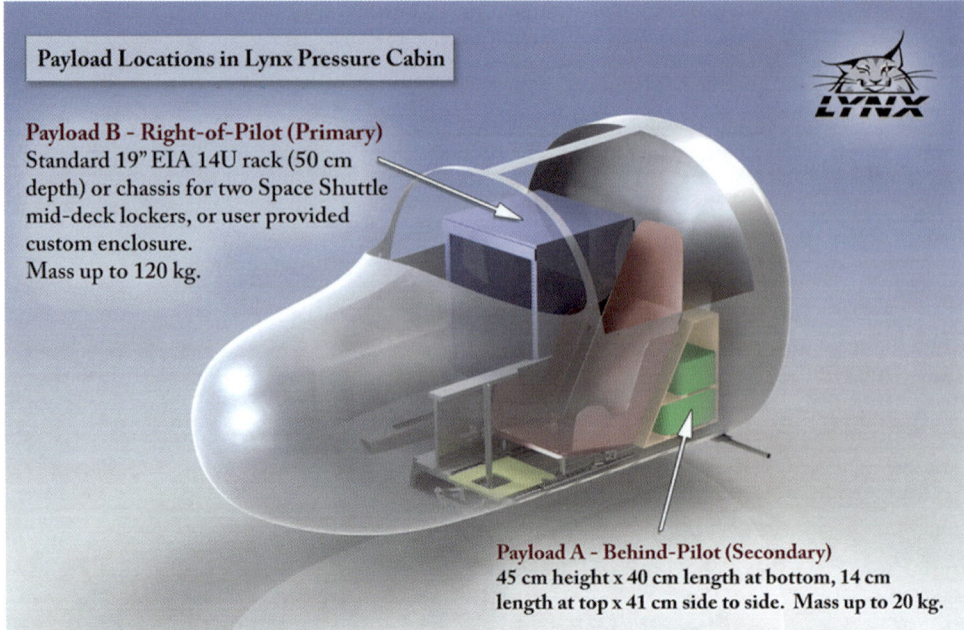

Payload Locations in Lynx Pressure Cabin

Payload B - Right-of-Pilot (Primary)
Standard 19" EIA 14U rack (50 cm
depth) or chassis for two Space Shuttle
mid-deck lockers, or user provided
custom enclosure.
Mass up to 120 kg.

Payload A - Behind-Pilot (Secondary)
45 cm height x 40 cm length at bottom, 14 cm
length at top x 41 cm side to side. Mass up to 20 kg.

8.1 Lynx Payload configuration. Courtesy: XCOR

SS2 Timeline

A typical flight to space on board SS2 will take about two hours from take-off to landing. The flight profiles (Figure 8.3) for astronaut flights and for autonomous payload flights will be the same. WhiteKnight2 (WK2) and SS2 will take off from Spaceport America at around 7 a.m. The mated vehicles will climb to approximately 15,000-meter altitude. There, after completing vehicle checks, SS2 will be released from WK2, free falling for about three seconds before igniting its rocket motor, which will boost the vehicle to a speed of about Mach 4. SS2 will coast upward, reaching a peak apogee as high as 110 kilometers.

During its coast phase, SS2 will "feather" its wings in preparation for re-entry. In SS2's feathered configuration, the entire tail structure is rotated upwards 60°, creating high drag as the vehicle enters the atmosphere. It's a method of re-entry that results in a lower skin temperature and smoother (i.e. more comfortable) G-loads during re-entry. Following re-entry, the tail resumes its original position and SS2 glides safely back to the runway. A typical timeline is:

Mated climb: ~60–90 minutes
Boost: ~60 seconds
Coast: ~240 seconds
Re-entry: ~80 seconds
De-feather: ~30 seconds
Glide to land: ~15 minutes

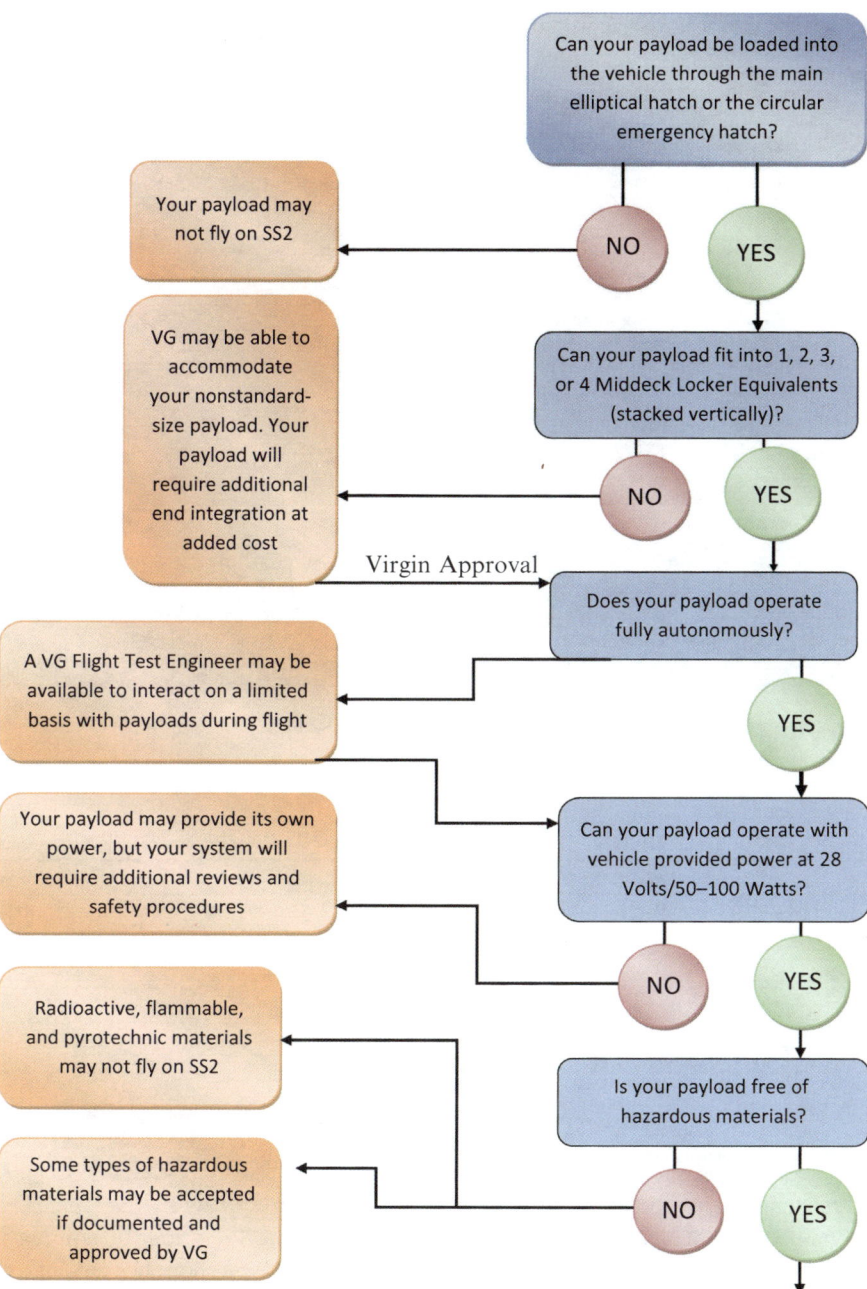

8.2 Payload User Guide. Courtesy: Virgin Galactic

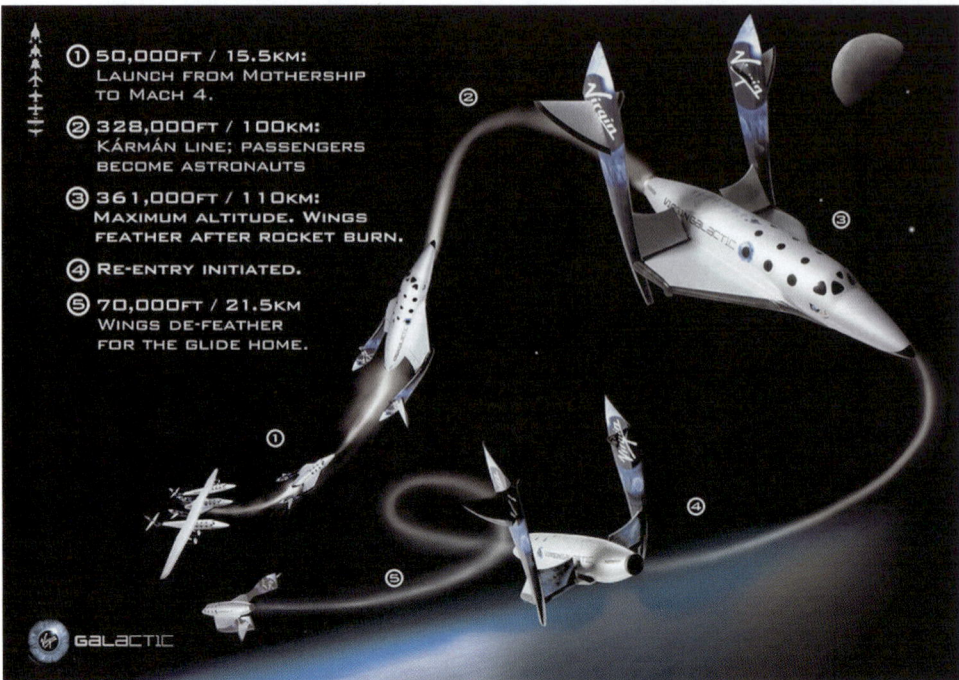

8.3   SpaceShipTwo flight profile. Courtesy: The SpaceShip Company/Virgin Galactic

Table 8.1.   Expected G-Loads for Flight and Crash Conditions Direction.

	Maximum boost loads	*Maximum re-entry loads*	*Crash loads*
Front/back (Gx)[1]	+0.1/–3.4	+1.4/–1.5	+15.8/–0.0
Left/right (Gy)[2]	+0.0/–0.0	+1.8/–1.8	+2.8/–2.8
Down/up (Gz)[3]	+3.7/–1.0	+8.4/–0.1	+4.5/–4.5

[1]Forward direction (eyeballs out)
[2]Sideways direction (eyeballs left/right)
[3]Up and down direction (eyeballs up/down)

Microgravity environment and G-loading

If you're planning to fly payloads on SS2, you don't have to worry too much about G-loading because the vehicle was designed to carry humans. So, as long as your payload is designed to operate under nominal flight loads (Table 8.1), you shouldn't encounter any problems. To ensure safety, your payload and its mounting hardware must be designed to withstand the nominal and off-nominal loads indicated in the table, which means the payload structure can't detach when subjected to crash loads.

Another factor to consider is the cabin humidity, temperature, and pressure environment. During flight, SS2 will be pressurized to 1,524-meter equivalent altitude (about the same as the latest breed of commercial airliners) and, while the vehicle is mated to WK2, the cabin will be monitored for pressure (12.2 psia), temperature (between 40 and 90°F), and humidity (less than 75%) to ensure a shirt-sleeve environment. Prior to release, SS2 will isolate itself from WK2 and switch to its dry-air pressurization system.

If you're a commercial astronaut, you will probably arrive at Spaceport America three days before launch to integrate your payload. Ideally, you should probably budget at least three days for training and another two for payload integration. SS2 contains a set of payload-mounting plates, which provide support and structure for all sorts of payloads. Each mounting plate was designed to accommodate a variety of payload containers, including NASA Shuttle Middeck Lockers, Cargo Transfer Bags, server racks, and CubeSats. If you happen to have your own container system, Virgin can provide you with a mounting system ... at an additional cost. One mounting plate can accommodate as many as four middeck lockers or containers, and each mounting plate can accommodate up to 90 kilograms, including containers.

On a standard science flight, SS2 will be outfitted with up to eight plates, with five on the port side and three on the starboard side, which contains the crew emergency egress hatch; this is where the Flight Test Engineer will sit to deal with any off-nominal situations. Another feature you should be familiar with is the bolt pattern on each payload rack. The bolt pattern is designed to accommodate the Middeck Locker but it can also be used with custom structures. Each of the 16 mounting holes you can see in Figure 8.4 is a sleeve bolt receptacle that will require the proper threaded insert on your payload attachment.

Obviously, to expedite payload integration, it makes sense to design your payload to fit within Virgin Galactic's payload specifications. Not only will this speed up your integration process, it will also keep your costs low. The standard SS2 flight configuration will utilize payload containers or mounting stations of three different sizes (Table 8.2). On a typical SS2 science flight, Virgin will offer six spots for Standard Payloads, two spots for Small Payloads, and three spots for Mini Payloads. If you want to avoid extra paperwork required for structural analysis, Virgin encourages payload users to use the NASA Shuttle Middeck Lockers or Cargo Transfer Bags.

Chances are your payload will need electrical power. No problem, because SS2 will offer a 28-volt power supply per payload location, with other power levels offered on a case-by-case basis. In addition to power requirements, payload users will likely require vehicle data and flight instrumentation through an Ethernet interface – a service that comes as standard for those flying on Virgin Galactic, since data from the Inertial Navigation System (INS), the Global Positioning System (GPS), and the Air Data System (ADS) can be provided to the payload (this includes several video feeds).

When it comes to actually fitting your payloads, be sure to make a note of the size of two available points of entry: an elliptical main hatch with a major diameter of ~84 centimeters and a minor diameter of ~66 centimeters, or the circular emergency exit hatch with a diameter of ~66 centimeters. Of course, if your payload doesn't fit through these openings, you always have the option of disassembling your payload and piecing it together inside the cabin, but Virgin Galactic would prefer to avoid the ship-in-the-bottle approach! As far as positioning your payloads are concerned, you can install them at locations in the left and right side of the cabin just as long as your payload doesn't impede egress.

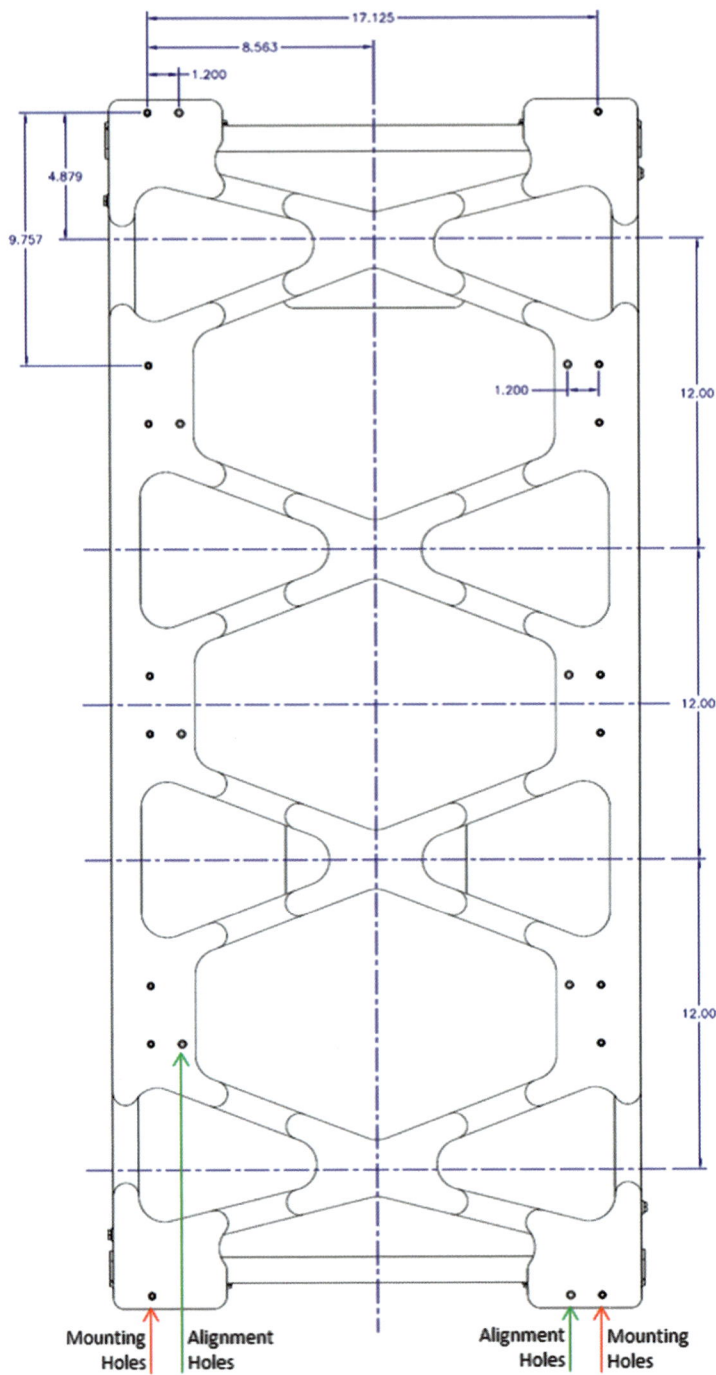

8.4 Sleeve bolt receptacles on SpaceShipTwo. Courtesy: The SpaceShip Company/Virgin Galactic

Table 8.2. Payload Specifications Type.

	Dimensions	*Volume (MLEs)*	*Max. weight* (lbm)*
Standard payload	18.50″ W × 46.50″ H × 21.50″ D	4	120
Small payload	18.50″ W × 23.00″ H × 21.50″ D	2	60
Mini payload	18.50″ W × 11.25″ H × 21.50″ D	1	30

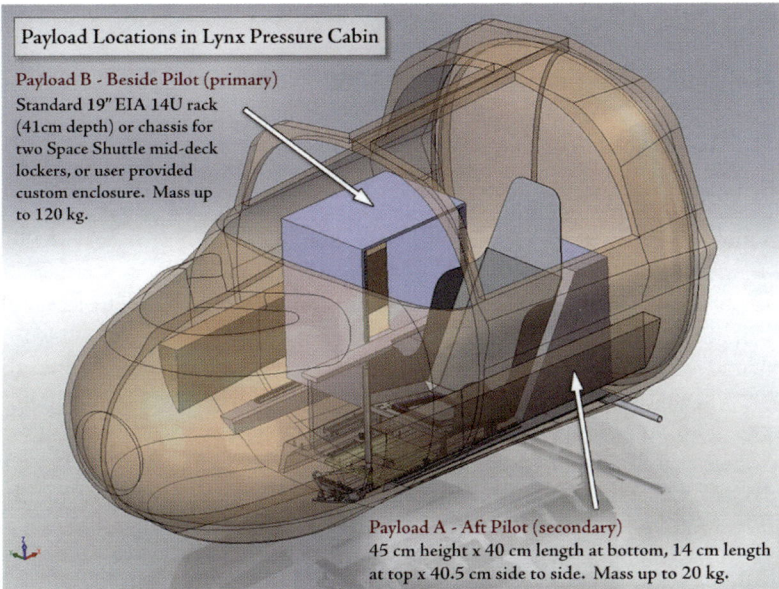

8.5 Payload location of Shuttle mid-deck lockers. Courtesy XCOR.

During a typical science flight, the SS2 cabin crew will consist of a pilot, co-pilot, and a Flight Test Engineer (FTE) appointed by Virgin Galactic. It's the FTE's job to monitor payloads for unusual behavior that could jeopardize the vehicle and/or crew. If a switch needs to be flicked or a button pressed, chances are the FTE can do that, but that's not in their job description so, if your payload requires more human interaction, it's probably best to either have a second Virgin Galactic FTE on board or fly yourself. Of course this becomes expensive!

Looking over the horizon, Virgin Galactic plans to offer additional payload-mounting options outside the pressurized cabin. Payloads in these locations (Figure 8.5) will be exposed to a similar environment as outside conditions. Virgin Galactic may also offer payload customers the ability to deploy payloads from the vehicle during flight.

By now, you may be wondering how much your payload will cost. Pricing for payload flights on SS2 will depend on the requirements of the payload but, generally speaking, pricing will be commensurate with pricing for astronaut flights, which means full flights will cost US$1.2 million before payload analysis and integration costs are factored in. Payloads occupying about the mass and volume of an astronaut and their seat will cost approximately the same as a single seat ticket price – US$250,000. If that's too much money, you can always go the XCOR route, although the Lynx doesn't have a roomy cabin to float around in.

FLYING A PAYLOAD WITH XCOR

The Lynx (Table 8.3) will offer a variety of multi-mission primary and secondary payload capabilities, including in-cockpit experiments, externally mounted experiments, astronaut training, upper atmospheric sampling, and microsatellite launch. Primary payloads pay for the flight, while secondary payloads are on a ride-share basis with a primary payload, typically for a nominal ride-share fee. As you can see in Figure 8.1, primary payloads will be located in the area to the right of the pilot on board the Mark II or, in the case of the Mark III, on top of the vehicle in an experiment pod. For the Mark II version, the primary internal payload will accommodate a maximum mass of 120 kilograms to suborbital altitude. In the event that the payload is a man-tended one, the right seat can be a human in a pressure suit. If not, the right seat can accommodate two stacked Space Shuttle middeck lockers (MDLs), or a standard 48-centimeter electronic equipment rack (Figure 8.5). For the Mark III vehicle, the primary payload location is an external dorsal mounted pod, which holds up to 650 kilograms and is large enough to hold a space telescope or a two-stage carrier to launch nanosatellites into low Earth orbit (LEO). This pod can also carry experiments that return with the vehicle.

The Lynx Mark I is a manned sub-suborbital vehicle that features a double-delta wing and twin outboard vertical tails. The vehicle is flown by one pilot and takes off and lands horizontally from a spaceport runway using a three-wheeled retractable landing gear. The airframe is composed of all composite materials with thermal protection added to the nose and leading edges. At just 8.5 meters in length and with a span of just 7.3 meters, the Lynx is the sports car of commercial suborbital vehicles, weighing less than twice the weight of a Cadillac Escalade. Although the Marks I, II, and III all look similar (they all have the same payload dimensions), there are subtle differences. For example, the Mark II weighs about 100 kilograms more than the Mark I and it can carry heavier payloads than its prototype cousin. The Mark III features a modification that allows it to carry a 650-kilogram payload on top of the vehicle in place of internally carried payload mass.

Table 8.3. Lynx Variants' Capabilities.

Service	Lynx Mark I	Lynx Mark II	Lynx Mark III
Payload destination	Sub-suborbital	Suborbital	Suborbital
Nominal vehicle altitude	83 km	103 km	103 km
Time in microgravity	60 sec	3 min	3 min

When operational, the Mark I will conform to the following flight profile: 13 seconds after engine start, the vehicle will take off, and about two and a half minutes after take-off, the engines will shut down, upon which the vehicle will coast upward for another 30 seconds before entering the low-acceleration period at 61-kilometer altitude. Microgravity time will be about one minute, after which the vehicle will enter the pull-out acceleration and complete maneuvers that will see it touch down about 30 minutes after take-off.

The Lynx Mark II flies the same flight profile as the Lynx Mark I, except the Mark II is under thrust for a longer time, reaches a higher apogee, and takes longer to return to the ground. After a 660-meter ground roll, the Mark II will take off and, three minutes into flight, the engines will shut down. The vehicle will enter its coast phase that will carry it to an apogee of 100 kilometers, after which the vehicle will follow the same profile as the Mark I. The Mark III, which has to be able to expose payloads to the vacuum or low-density air of the upper atmosphere, has to reach a higher apogee than the Mark II, so it is under thrust for a longer time and takes longer to return to the ground.

To maneuver for payload positioning when aerodynamic Q is high, the pilot positions the vehicle by manually activating the Lynx's power-assisted elevon and rudder controls, trim flaps, and drag brakes. When aerodynamic Q is low, the pilot manually controls pitch, yaw, and roll with two independent reaction control systems (RCS). Using this system allows the pilot to orient the vehicle in any direction and stabilize within 10 seconds. Like I said, it handles like a sports car.

The Lynx can carry in-cabin and external payloads, as well as a commercial astronaut or tourist. But, for a payload mission, XCOR distinguishes between primary and secondary payloads; primary payloads determine the flight trajectory, date, and mission objectives, while secondary payloads are manifested with a primary payload, which may be a spaceflight participant, and do not control the flight date, trajectory, or mission objectives. The Lynx offers four payload integration locations: two in the cabin and two outside (unpressurized). Just like Virgin Galactic, those wishing to fly payloads must conform to XCOR requirements (Table 8.4) for size, strength, containment, vehicle safety, and secure integration.

Payload A is latched to the seat track aft of the pilot's seat and in front of the pressure dome. The shape is a right-triangular volume with the top cut off (approximately the width of the pilot's seat). The only crew access to this payload during flight is a payload on/off switch on the pilot's instrument panel. *Payload B* will be located in the position normally occupied by the spaceflight participant (SFP)'s seat. The right seat can be removed and an experiment rack payload container that stows two Space Shuttle middeck lockers can be placed here. *Payloads* cowling port (*CP*) and cowling starboard (*CS*): two cowling payloads can be carried on each flight on each side of the vehicle in the aft cowling, external to the cabin. Payloads CP and CS each hold a double CubeSat or a cylindrical volume that encloses two CubeSats stacked end to end. *Payload D* is the dorsal pod mounted on top of the Lynx. XCOR will offer a small and a large pod option. No environmental controls are provided to this external payload. Payload D can be launched using a payload-supplied spring sized to payload requirements, by gas pressurization supplied by the vehicle, or by an upper stage developed by XCOR.

Table 8.4. XCOR Payload Locations.

Location	In-cabin or external?	Lynx Marks I and II	Lynx Mark II	Payload limits
Payload A	In-cabin	20 kg maximum	20 kg maximum	No limitations on other payloads
Aft of pilot		50 cm height × 40.5 cm width; bottom depth: 46 cm; top depth: 16.5 cm	50 cm height × 40.5 cm width; bottom depth: 46 cm; top depth: 16.5 cm	
Payload B	In-cabin	120 kg maximum	120 kg maximum	Precludes carrying a SFP
Beside pilot (primary)		2 Shuttle middeck lockers	2 Shuttle middeck lockers	
Payload CP and CS	External	1.5 kg each	1.5 kg each	No limitations on other payloads
Cowling – port and starboard (secondary)		15 cm diameter × 20 cm depth (fits a double CubeSat)	15 cm diameter × 20 cm depth (fits a double CubeSat)	
Payload D	External	280 kg maximum (Mark I only)	650 kg maximum at second-stage ignition	Kilogram for kilogram, this payload reduces the allowable mass for all other payloads
Dorsal pod (primary)		Cylindrical volume: 43 cm diameter × 240 cm length (155 cm full diameter length)	Cylindrical volume: 76 cm diameter × 340 cm length	

If your payload happens to be an animal experiment and you're concerned about how your test subjects will fare during the flight, you don't have to worry … not too much. The cabin will be noisy during take-off (similar to the noise experienced inside a single-engine Cessna) and it will be pressurized to 2,700 meters and temperature controlled to about 20°C. Having said that, you should be aware that Spaceport America experiences some pretty dramatic temperature swings, and it can be quite cold early in the morning. There is also the possibility of a rapid or an explosive decompression; while this is also a possibility in a general aviation flight, such an event occurring during the microgravity phase of the flight could be deadly for any unprotected animals. When XCOR made its payload presentation at the 2013 Next Generation Suborbital Researcher's (NSRC) conference, there were a couple of questions about radiation shielding. The Lynx doesn't have radiation shielding for the very simple reason that it doesn't need any, although shielding can be provided at the payload user's expense. Vibration shouldn't be a problem because the Lynx doesn't use solid or hybrid rocket motors, which means the shock and vibration environment is milder than for a piston-propeller-driven airplane. As for acceleration loads, XCOR predicts G-forces of 3.5–4 G in the dive and a worst-case emergency pull-out envelope of +8/–6 G.

Another issue for those flying payloads is positioning accuracy and stabilization control. Thanks to the Lynx's RCS, this shouldn't be a problem because the pilot can maneuver the Lynx into whatever position is required to meet payload requirements. During operation of the RCS, payloads will be subject to a rotation of up to 15° per second[2] and the positioning and rate accuracy should be to within a couple of degrees or less in each axis.

When you're designing your payload for a flight on board the Lynx, there are a few requirements XCOR would like you to be aware of. One of these is the pre- and post-flight environment which, in the area around Spaceport America, can include plenty of sun, high winds, and flying dust and sand. To avoid your payload being affected by adverse environmental conditions, it should be designed to survive all aspects of the area's climate. Your payload also needs to be designed to be secured to the structure of the Lynx vehicle to withstand all forces in flight, including a pilot-survivable crash, which means you have to provide XCOR with documents that describe the structural materials of your payload and their strengths. To make the whole process of payload acceptance a little easier, it's probably best to avoid trying to fly payloads containing hazardous liquids that can release toxic substances into the cabin in the event of a crash or explosive decompression. On the subject of crashing, you also need to design your payload with a built-in capability that safely shuts down or fails without leaking liquids, emitting flames or toxic gases, or otherwise having any impact outside the payload container. Another safety consideration is electromagnetic interference (EMI). If you read the XCOR payload user's guide, you'll note that XCOR reserves the right to power down any payload determined to be interfering with other experiments or aircraft instrumentation systems, so it's important to factor this into your design. Finally, there is the issue of paperwork that you must submit before your payload is accepted; a payload quick reference sheet; a hazard source checklist; ground-support requirements; proof of liability insurance; final payment. Once your payload has been accepted, you can look forward to the flight, the timeline of which will follow the activities listed in Table 8.5.

If you're interested in flying a payload on board the Lynx, XCOR suggest contacting them at least 90 days before the flight date. This gives you and XCOR time to transfer your experiment into an approved payload container and complete the required paperwork, insurance, and cleared payment; incidentally, XCOR must receive the completed payload documentation, proof of insurance, and proof of payment no later than 30 days before flight. To fly a payload on the Lynx,[1] be prepared to provide the following:

- Principal Investigator – name, contact information, and sponsoring institution
- Description of payload and mission goals
- Number of flights and schedule

---

[1]   Either e-mail or call Andrew Nelson (e-mail: anelson@xcor.com, phone: 661-824-4714).

Table 8.5. Lynx Preflight Timeline.

Timeline	Activity	Details
150 minutes before flight	Load, secure, and check payloads	Payloads will be loaded and secured in their locations prior to the pilot's briefing. All connections and data-gathering equipment will be tested prior to the vehicle being towed for fueling
120 minutes before flight	Final payload checks	
	Preflight briefing	Experimenter's attendance at the briefing is optional and limited to observation. All payload information provided to XCOR must have been previously documented
30 minutes before flight	Crew ingress	Crew enters cabin
0	Take-off	
During flight		Operate payload
30	Landing	
60	Return to hangar	Deplane. Remove payloads. Check for vehicle and payload damage
120	Post-flight debrief	

- Estimated size and weight of payload
- Payload integration location(s)
- Preliminary hazard analysis and planned controls
- Payload services required from XCOR; manual payload actuation, payload deployment and launch, electrical power, voice and data communication, data recording and telemetry, environmental control, positioning and stabilization flight maneuvers, cryogens, or overboard venting
- Ground support required, including work space and storage, hours of access
- Special payload handling; payload refrigeration, live experiments, and confidentiality requests

FLYING A PAYLOAD WITH BLUE ORIGIN

One unexpected presenter at the 2013 NSRC was Blue Origin. The company, whose secretive approach to doing business is probably the envy of the National Security Agency (NSA), provided attendees with an update on the company's developments and also an overview of how to fly payloads on board their New Shepard vehicle. Blue Origin's first goal is to develop a fully reusable vertical-take-off and vertical-landing (VTVL) suborbital system. After the company has a suborbital vehicle, it plans to develop an orbital spacecraft that uses the suborbital vehicle as the upper stage of the orbital system. The Blue Origin mantra is an incremental development approach that focuses on regular flight testing – an approach that has been accelerated thanks to NASA collaboration. The plan is to fly payload specialists with experiments, although New Shepard isn't a NASA Flight

Table 8.6. Cabin Payload Bays.

Accommodation	Single size	Double size
Experiment volume	0.0473 m³	0.102 m³
Approximate interior dimensions	51.13×41.78×22.65 cm (less hinge area)	51.13×41.78×48.94 cm (less hinge area)
Experiment mass	11.34 kg	22.68 kg
Power	28 V DC	
Data recording	Post-flight download with synchronized trajectory parameter measurements	
Communications	Low data rate link for experiment telemetry and control	
Experiment configuration software	Blue Origin's *REMConfig* software configures experiment command sequence, such as turning on sensors when microgravity is reached	

Opportunities platform at the time of writing. Blue Origin doesn't know when New Shepard will fly and the company isn't marketing passenger flights. The company has just three science payloads scheduled and, when they are fully operational, they are aiming for one flight per day.

The New Shepard vehicle consists of a pressurized crew capsule carrying experiments perched on top of a propulsion module. New Shepard will launch from the company's West Texas launch site, accelerating for approximately two and a half minutes before the propulsion module shuts off its engines and coasts into space, where the crew capsule will separate from the propulsion module. The two vehicles will re-enter and land separately near the launch site, the propulsion module autonomously performing a rocket-powered vertical landing while the crew capsule will land under canopy.

If you're interested in flying your experiment on board New Shepard, Blue Origin has developed a payload bay that comes in two standard sizes (Table 8.6). Researchers install their experiments inside each cabin payload bay and several bays are grouped and mounted together into a rack on top of an avionics cabin payload bay that supplies experiment power, data processing, and data storage for the flight. As with Virgin Galactic and XCOR, Blue Origin's research accommodations will be fairly flexible and researchers will be able to provide their own racks by special arrangement.

THE SUBORBITAL PAYLOAD AGENT

If you happen to be a scientist who would like to fly your experiment in space but don't have the time or the resources to jump through all the administrative hoops required by the operators, what do you do? Well, one solution – which isn't available yet – is for a third-party organization to arrange the science flight. These "payload agents" would liaise with the scientists and operators, handle the technical details of integrating the experiments on the vehicles, and ensure the experiments on each flight are complimentary. But why can't the scientists use the travel agencies the suborbital companies are using to interface with customers and coordinate sales activities that way? Well, these travel agents won't be up to the task because the payload customer base is focused on scientific research and not human entertainment. Of course, there is the option of developing the sales/customer service talent in-house, but this would be expensive and distracting from the operator's focus of flying rockets. So, operators launching payload and scientists will probably need their own class of travel agent, their own out-sourceable sales force, which would be a Suborbital Payload Agent (SPA).

The SPA would be responsible for interfacing with the customers (scientists who want to accompany their payload or those who simply want their payload flown on a flight rack), integrating experiments into flight racks, and delivering integrated flight racks to launch operators for flight. After each flight, the SPA would deliver payloads and data back to the customer. Basically, all the coordinating of payloads, launch logistics, and flight payment would be done by the SPA. A SPA would be a huge benefit to the customer, not only because it would relieve them of all the paperwork we mentioned before, but it would also create more flight opportunities. How? Well, imagine if a customer wants to fly on a particular flight but the manifest is full of tourists. Not a problem, because the SPA will have signed agreements with multiple launch operators, so, in the event of a full manifest or, God forbid, a crash, the payload would simply be switched to alternate providers as required. Also, as the commercial suborbital market becomes more competitive, the SPA could leverage their buying power when purchasing flights from the launch provider and the savings could be passed on to the customer.

The SPA would be beneficial to the launch providers too, because it would allow them to focus on their core competency – launch operations. Also, outsourcing sales and customer service responsibilities and the associated costs time and money, so the SPA would be a godsend and, even if an operator decides to develop a sales force of their own, an SPA would still allow an operator to increase demand at a lower cost.

9

How to Get There

Buying a ticket to space – even suborbital space – is expensive. So what options are there for getting your human-tended payload – you! – into space? If you happen to be a wealthy individual, you can simply hand over your cash to an accredited space travel agent, but if you're a scientist, chances are your options are more limited. This chapter explains some of the strategies you can employ to fly in space and offers some last-minute considerations.

The explosion in private wealth over the past few years means a surprisingly large number of people can afford US$250,000 Virgin Galactic flights. To get some idea of how viable the suborbital spaceflight business might be, the Federal Aviation Administration (FAA) conducted a study and concluded that more than 1,000 people a year would likely purchase suborbital space tours – that's US$250 million in revenue per year. Just for Virgin Galactic. The study also estimated about 80% of demand for suborbital flights would come from wealthy individuals interested in space tourism, while business, research, and government would account for about 20%, which is great news for those who don't have a million-dollar-plus annual income. Another interesting and positive statistic for those in the business of launching people into space is that, globally, there are 11 million high-net-worth individuals with over US$1 million in liquid assets (according to consulting firm Capgemini). Some of these wealthy individuals have already bought tickets. They include celebrities Paris Hilton, Angelina Jolie, Ashton Kutcher, Justin Bieber, and Tom Hanks, who got a taste for weightlessness while filming *Apollo 13*. The jet-set list also includes retired Formula One aces Michael Schumacher, Rubens Barrichello, and Niki Lauda, as well as an assortment of wealthy businessmen such as real estate magnate Ashish Thakkar and Danish investment banker Per Wimmer. If you would like to join them, just visit the Virgin Galactic website and click "Booking" or visit an accredited space agent:

"We need affordable space travel to inspire our youth, to let them know that they can experience their dreams, can set significant goals and be in a position to lead all of us to future progress in exploration."

Burt Rutan, Founder, Chief Technical Officer

E. Seedhouse, *Suborbital: Industry at the Edge of Space*, Springer Praxis Books, DOI 10.1007/978-3-319-03485-0_9, © Springer International Publishing Switzerland 2014

If you decide to use the online route, you will be asked to complete a booking form and one of Virgin Galactic's Astronaut Relations team-members will get back to you within 24 hours. If you decide to book through a space agent[1] (there are about 120 of them dotted around the world and the number is growing), then simply click on the world map on the Virgin website to see where the closest one is to you. Don't have the US$250,000? No problem. A US$25,000 deposit will hold your seat and you can pay the rest when you get your launch date.

Not surprisingly, California, with its high density of tech entrepreneurs, is a hotbed for selling suborbital tickets. In fact, about half the people who have signed up for Virgin Galactic are tech CEOs or entrepreneurs and, because they enjoy Pioneer Astronaut status, they have exclusive access to Galactic events. These events include everything from astronaut forums with Sir Richard Branson on his Caribbean island to opportunities to tour Scaled Composites to watch spaceships being built. Pioneer Astronauts have also attended events hosted by Branson at South African game reserves, at estates in the Atlas Mountains in Morocco, and stayed in the Ice Hotel in the Arctic, location of Spaceport Sweden.

Having spent US$250,000, you probably want some assurances about the safety of your flight. A few years ago, Branson offered William Shatner, the actor who portrayed Captain Kirk in the *Star Trek* series, a free ride into space on the inaugural launch of the VSS *Enterprise*, but the iconic actor turned it down, saying "I do want to go up but I need guarantees I'll definitely come back". Well, there are no guarantees in the space business, but SpaceShipTwo (SS2) is about as safe a vehicle as they get. First of all, SS2 (Figure 9.1) launches horizontally from an aircraft at around a 13,000-meter altitude rather than vertically from the ground. Why is this safer? Well, ground launches are more dangerous than air launches because the vehicle has to pass through the denser regions of the atmosphere and, to do this, the rocket motor's exhaust has to work very hard to get the spacecraft moving at the high speeds necessary to get through the thicker air. The problem with this is that traveling at very high speeds in the lower atmosphere creates a great deal of drag, which in turn produces high structural loads and needs a stronger and heavier fuselage. It also means large quantities of fuel are required for the longer-duration burn, which means an even bigger fuselage is required, which leads to even more weight, which means you need even more fuel to lift the extra weight. This way of launching rockets also means everything has to go right the first time; if it doesn't, there aren't many options.

Burt Rutan, genius that he is, reckoned the safest and most efficient strategy was to air-launch SS2. This strategy meant the rocket motor only had to burn for a very short time to reach space and, if there were any problems during the boost phase, the rocket motor could simply be shut down and the spaceship would return as a glider. In addition to flying a safe and proven (remember, the X-15 planes flew air-launched missions to space) flight profile, future astronauts can also take comfort knowing SS2 is a structurally sound vehicle; built of carbon fiber composite, which is four times the strength of steel and a quarter of its weight, SS2 is not only very light, but also very strong.

---

[1] As part of Virgin Galactic's training program, these unique travel agents meet with test pilots and visit New Mexico to tour Spaceport America. Every month, they receive a call from Virgin Galactic to report on their progress. Sometimes, there isn't much to report, although once revenue flights start taking off, business should pick up.

9.1 SpaceShipTwo slung under its mother ship, WhiteKnightTwo. Courtesy: Virgin Galactic/
The SpaceShip Company

Safety isn't just found in the flight profile and the construction materials. The way this inspirational spaceship gets back to Earth also provides its occupants with another layer of security thanks to Rutan's radical re-entry method originally utilized by SpaceShipOne (SS1). Re-entry has long been considered one of the most challenging and dangerous phases of a space mission and Rutan was determined to find as safe a solution as possible by employing the Scaled Composite's mantra of safety through simplicity. His solution was "feathered re-entry". Very simply, Rutan designed SS2's pivoted wings to act like a shuttlecock, to slow and control the spaceship's re-entry.

Once SS2 is out of the atmosphere, its tail can be rotated upwards to about 65° – a configuration that allows an automatic control of attitude with the fuselage parallel to the horizon. The configuration also creates very high drag as SS2 descends through the upper atmosphere and, because the feather configuration is so stable, the pilot can almost fly hands-free. Yet another safety feature stemming from the combination of high drag and low weight (due to the composite materials used to build SS2) is the low skin temperature during re-entry, rendering thermal protection systems such as heat shields or tiles redundant.

So safety shouldn't be a concern. But what about room to perform those aerobatics that you see astronauts doing on board the International Space Station (ISS)? And how easy will it be to take photos? The good news is SS2 is roomy. Perhaps not as spacious as the ISS, but definitely roomy. For those who like numbers, the cabin is more than six meters long and more than two meters in diameter, which is about the same volume as the interior of a Falcon 900 business jet, which several Virgin Galactic passengers will no doubt be familiar with. Taking photos won't be problem because each passenger is assigned a seating position with two large windows – one side window and one overhead window (each 33 centimeters wide and 43 centimeters high).

STRATEGIES TO FLY A MISSION

If you don't happen to be independently wealthy but still have a powerful urge to fly in space, what can you do? Well, first you have to be patient because, as I'm writing this, revenue flights are still on the horizon, as those who attended the 2013 Next Generation Suborbital Researchers Conference (NSRC) were reminded. On the first day of the annual event, attendees gathered in the packed Omni Interlocken hotel ballroom to discuss their dreams for commercial spaceflight, as they had done for the past three years. In 2013, the mantra was the same as it had been in 2010, 2011, *and* 2012: one day soon, commercial spacecraft will take off on a daily basis, carrying space tourists and scientist astronauts. Not surprisingly, some attendees recalled similar proclamations being made at the first NSRC. Back then, it was predicted that "by the end of 2011 or beginning of 2012 you're going to see spaceports struggling to deal with a flight rate that's completely unprecedented". Fast forward to 2013 and there are eight FAA-licensed US spaceports, but very little action. What's the hold-up?

   "It takes a while in the space business," said Alan Stern, an associate vice president at the Southwest Research Institute (SwRI) in Boulder, Colorado, and a driving force behind NSRC. As with so many endeavors, the reason this suborbital spaceflight business is taking a while is mainly due to funding problems. NASA has provided much of the seed money to give commercial spaceflight a boost, because the agency sees it as cheap access to space. But, at NSRC 2013, NASA deputy administrator Lori Garver highlighted some of the ways NASA had fallen short of its goals. Three years ago, Garver had promised commercial spaceflight US$15 million annually, but fiscal congressional restraint reduced that to about US$10 million. NASA also combines its suborbital research program with payloads flown on balloons, sounding rockets, and parabolic flights, which means there is extra competition for limited dollars. This Flight Opportunities Program has selected seven vendors for commercial suborbital flights but, as Garver admitted at the conference, "We all wish things were coming around even faster". Unfortunately, if the US State Department has its way, progress may even more stifled. That's because the US State Department has proposed adding man-rated suborbital spacecraft to the US Munitions List (USML – see Chapter 6), which includes sensitive technologies such as military equipment and nuclear weapons. It's all part of President Obama's Export Control Reform effort and, if it passes, it means trying to ferry a Lynx or an SS2 to another country would prove very, very difficult.

BECOME A CITIZEN-ASTRONAUT

The news wasn't what some impatient suborbital researchers wanted to hear, but Alan, as always, was optimistic. "The world is about to change very rapidly as these vehicles come online." I have no doubt Alan is right and most of those at NSRC shared his optimism, among them Citizens in Space (www.citizensinspace.org) founder Ed Wright. Ed, who is chairman of the United States Rocket Academy and project manager for Citizens in Space, previously worked in the computer software industry and was president of X-Rocket, LLC, which operated a high-performance MiG-21 jet. Citizens in Space, a Universities

9.2   XCOR's Cub Carrier. Courtesy: XCOR

Space Research Association (USRA) project, has already acquired a contract for 10 suborbital spaceflights with XCOR (the largest single bulk purchase of suborbital flights), and plan to offer the tickets to citizen scientists and researchers who play by their open-source rules. The organization has already selected three astronaut candidates who are already in training. In partnership with USRA, the organization plans to conduct Pathfinder Astronaut Training Workshops, to beta-test the citizen-astronaut training activities they are developing. The organization is also developing the Lynx Cub Payload Carrier (Figure 9.2) – a unit that can hold up to a dozen small experiments in a space behind the pilot's seat. The carrier means an astronaut can sit in the passenger seat while also carrying several scientific experiments as secondary payloads.

Another initiative dreamt up by the organization is the Space Hacker event, the first of which took place in May 2013. Sponsored by the Mountain View-based Silicon Valley Space Center, the event kicked off with Ed Wright talking about Citizens in Space and the organization's High Altitude Astrobiology Challenge, which offered up to US$10,000 cash for the best space organism-collection project. Khaki Rodway, XCOR's director of payload sales and operations, gave an overview of Lynx's capabilities and an update on the assembly of the first vehicle. During the two-day workshop, participants heard talks about

9.3   Water bears are a candidate for suborbital life sciences flights. Courtesy: NASA

experiments that will be carried out in the microgravity environment: everything from materials processing and protein crystal growth to fluid dynamics and water bears – tiny organisms that are virtually impervious to heat, radiation and vacuum conditions. That last experiment was suggested by NASA astronaut Yvonne Cagle, who suggested flying water bears (Figure 9.3) into space to study how humans adapt to weightlessness.

Citizens in Space emphasizes they are only interested in experiments that advance the fields of science or engineering and requires that hardware designs for the experiments be made available to other citizen scientists. While Ed acknowledges some of the experiments might not be as sophisticated as those submitted by well-funded university researchers, he sees promise in citizen science because it can operate outside the traditional spheres of research. For example, if you want an experiment to be funded by NASA and flown on the ISS, chances are your project has to align with a specific mission the agency has in mind. And, assuming you have an idea for such an experiment, you then have to wade through a paper blizzard of form-filling, submit to a lengthy peer-review process, and wait for the wheels of federal funding to turn. I'm not saying there is anything wrong with that process; it's just that many researchers would like to have access to a more streamlined and unencumbered process. And that's exactly what Citizens in Space offers.

FLY A PAYLOAD

If you would still like to go the federal funding route, the good news is that NASA has a program that funds experiments. More of a key technology development pipeline link, the Flight Opportunities Program calls for payloads through an announcement of opportunity.

In June 2013, the agency announced it had selected 21 space technology payloads for flights on commercial reusable launch vehicles, balloons, and a commercial parabolic aircraft. The selection represented the sixth cycle of the program, which has now facilitated more than 100 technologies with test flights – everything from systems that support cubesats to new sensor technology for planetary exploration. Of the 21 payloads selected in 2013, 14 will ride on parabolic aircraft flights, two will fly on suborbital reusable launch vehicles (sRLVs), three will ride on high-altitude balloons, one will fly on a parabolic flight and a suborbital launch vehicle, and another will fly on a sRLV and a high-altitude balloon platform. Although most of the payloads that have been selected to date are not suborbital, the program acknowledges the impending arrival of suborbital revenue flights and, once Virgin Galactic and XCOR start flying regularly, more payloads will be flown on sRLVs.

The main goal of the program is to develop and mature new technologies (Technology Readiness Level 4+). Selected proposals are offered a flight (or multiple flights) on a sRLV/parabolic aircraft, but are not provided with funding for payload development. The program funds more than a third of the proposals received and information about how to apply can be found at the website https://flightopportunities.nasa.gov/.

A similar program to the Flight Opportunities Program is NASA's Game Changing Development Program, which supports research from academia, industry, and governmental agencies. The program funds researchers to take their technology from a proof of concept stage (TRL 3+) to the component testing phase in an applicable environment. The program generally provides funding ranging from US$125,000 to US$500,000 for payload development in preparation for demonstration flights on sRLVs but does not guarantee flights for selected proposals; the next step for payload developers is to propose for a flight through NASA's Flight Opportunities Program.

If space and Earth sciences is your specialty, you might be interested in NASA's Research Opportunities in Space and Earth Sciences (ROSES) Program, which accepts proposals related to various Earth and space science initiatives within NASA. Proposals submitted to this program must be related to one or more of the following NASA Research programs: Heliophysics, Astrophysics, Planetary Science, and Earth Science. The ROSES program is open to groups including government agencies, private organizations, and non-profits, and funding ranges from US$100,000 to US$1,000,000 per year for a period of up to five years.

WIN A TICKET

If you don't happen to be a scientist with access to research dollars and you don't have a deep wallet, how about entering a competition? That's what thousands of budding suborbital adventurers did at the beginning of 2013 when AXE created its Apollo Space Academy (AASA). AXE, a Unilever-manufactured brand leader in men's care, wanted to give its customers a chance to experience an adventure unlike any other, so they decided to hold a competition to win tickets to go on board the Lynx. As part of the biggest product launch in its 30-year history, AXE asked guys and girls from more than 75 countries to sign up for the AASA by creating an astronaut profile on AXEApollo.com and telling the world why they deserved to go to space. Top-voted candidates qualified for a challenge in their country, with the finalists winning a place in the final stage at the AXE Global Space Camp in Orlando, Florida, where 22 space travelers were selected based on competitive space-simulation challenges. Helping the company advertise the competition was veteran Moon walker, Buzz Aldrin, who said "Space travel for everyone is the next frontier in the human experience. I'm thrilled that AXE is giving the young people of today such an extraordinary opportunity to experience some of what I've encountered in space".

Immediately after the competition was announced, tens of thousands of budding suborbital adventurers posted videos on the AXE website, and then went to work drumming up support of social media sites to persuade/coerce family and friends to vote for them. One finalist was Tale Sundlisaeter, who was the Norwegian winner. Tale, who at the time of the competition was a researcher and writer for Norway's technical weekly magazine, Teknisk Ukeblad, soon generated a strong following and garnered thousands of votes with the result that the Royal Norwegian Air Force offered her a flight in an F-18 (Figure 9.4) to help her momentum. In July, she got the call saying she had won her ticket to Space Camp. In December 2013 she, along with 21 other winners, won the prize of a ticket to space and the possibility of becoming Norway's first astronaut.

PERFORM RESEARCH

If your research institution has deep pockets, perhaps you can buy a ticket to perform research, which is what the SwRI did when it signed the world's first commercial contract to send researchers to space. As part of the contract that was announced in 2012, SwRI made full deposits for two researchers to fly on Virgin Galactic's SS2, with the intent to make similar arrangements for an additional six seats for a total value of US$1.6 million. In addition to flying its own researchers, who will conduct scientific experiments developed by its in-house technical staff, SwRI also aims to assist American researchers who do not have direct spaceflight experience to develop and fly their payloads and personnel on suborbital missions.

The deal was dreamt up by SwRI's Dr. Alan Stern, Associate Vice President of SwRI's Space Division and former NASA Associate Administrator for Science. Stern and two of his SwRI colleagues, Daniel Durda and Cathy Olkin, have already started training for their suborbital flights, on which they plan to carry out three experiments: one involves monitoring the researchers' blood pressure, a second involves monitoring a box of rocks to find out how rubble behaves in the low gravity of asteroids, and the third will test the performance of an ultraviolet imager that could be used to examine planets and other objects at wavelengths that are blocked by Earth's atmosphere. In addition to the Virgin Galactic deal, SwRI also inked a deal with XCOR Aerospace for six flights on board the Lynx, with options for three more.

9.4 (**a**) Tale Sundlisaeter in the back seat of an F-16. Courtesy: Tale Sundlisaeter. (**b**) Tale as she looks without fighter pilot gear. Courtesy: Tale Sundlisaeter

Appendix I

TO BOLDLY GO WHERE NO ANIMAL HAS GONE BEFORE – A SHORT HISTORY OF SUBORBITAL ANIMAL ASTRONAUTS

When you buy your suborbital space ticket, spare a thought for the animals who made your flight possible. Many years ago, before astronauts risked their lives, it was thought humans might not be able to survive the trip to space and back. So scientists launched animals – monkeys, chimps, and dogs mainly – to make sure they could launch them into space and bring them back alive.

On June 14th, 1949, a V-2 flight carrying an Air Force Aeromedical Laboratory monkey, Albert II, attained an altitude of 133 kilometers. Unfortunately, Albert II died on impact. On August 31st, 1948, another V-2 was launched carrying a mouse that survived impact. Then, on December 12th, 1949, the final V-2 monkey flight was launched at White Sands, carrying Albert IV, a rhesus monkey wired up to monitoring instruments. It was a successful flight, with no ill effects on the monkey … until impact; unfortunately, Albert IV didn't survive.

The Soviets, who also had manned spaceflight plans, kept a close eye on what the US was doing with their V-2 project, and decided they should conduct their own research. Basing their experiments on American biomedical research, Soviet rocket pioneer Sergei Korolev, his biomedical expert Vladimir Yazdovsky, and a small team of scientists used mice, rats, and rabbits as one-way guinea-pigs for their early tests. To gather data to design a cabin to carry a cosmonaut into space, they chose small dogs; the Soviets reckoned dogs would be less fidgety in flight than monkeys. Between 1951 and 1952, the Soviet R-1 rockets carried nine dogs, with three dogs flying twice. Each flight carried a pair of dogs in hermetically sealed containers that were recovered by parachute.

On July 22nd, 1951, Dezik and Tsygan (Figure A1) made spaceflight history when they became the first dogs to make a suborbital flight. Both dogs were recovered unharmed after traveling to an altitude of 110 kilometers. Dezik made another suborbital flight in September 1951 with a dog named Lisa, but, sadly, neither survived when their parachute failed to deploy. After Dezik's death, Tsygan was adopted as a pet by Soviet physicist

E. Seedhouse, *Suborbital: Industry at the Edge of Space*, Springer Praxis Books, DOI 10.1007/978-3-319-03485-0, © Springer International Publishing Switzerland 2014

A1.   Dezik and Tsygan were the first dogs to make a suborbital flight on July 22nd, 1951.
Courtesy: www.wikipedia.org

Anatoli Blagonravov. Tsygan never flew again and lived to old age. The Soviets continued their suborbital animal flights through 1960, but unfortunately not all of them had happy endings.

And then there was Ham who, on January 31st, 1961, became the first chimpanzee in space, aboard the Mercury-Redstone rocket on a suborbital flight very similar to Alan Shepard's. Ham's original flight plan called for an altitude of 185 kilometers and a speed of about 7,000 kilometers per hour. Unfortunately, due to technical gremlins, Ham's spacecraft reached an altitude of 260 kilometers and a speed of 9,600 kilometers per hour! The result was a landing 680 kilometers downrange rather than the anticipated 467 kilometers. During his flight, Ham experienced six and a half minutes of weightlessness. A post-flight medical examination found the well-traveled chimp to be slightly fatigued and dehydrated, but in good shape otherwise. Following a thorough medical examination, Ham was placed on display at the Washington Zoo in 1963, where he lived alone until September 25th, 1980. He was then moved to the North Carolina Zoological Park in Asheboro. Upon his death on January 17th, 1983, Ham's skeleton was retained for examination by the Armed Forces Institute of Pathology. His other remains were respectfully laid to rest in front of the International Space Hall of Fame in Alamogordo, New Mexico.

ASTROCATS – THE STORY OF FÉLIX AND FÉLICETTE

The American and the Soviets weren't the only countries sending animals into space; in 1963, the French government had a small team of cats undergoing intensive training for possible spaceflight, including compression chamber and centrifuge training. By all accounts, these feline astronauts-in-training don't seem to have suffered too much because 10 were deselected for overeating!

Of the lucky few who made the feline astronaut grade, Félix was the one chosen to undertake the first mission. Félix was a Paris street cat, although one report states he was bought by the French government from a dealer. Perhaps Félix didn't fancy being launched into space because he managed to escape, and was replaced at the last minute by a female cat, Félicette. So, on October 18th, 1963, it was Félicette who blasted off in a special capsule on top of a French Véronique AG1 rocket from the Colomb Bacar rocket base at Hammaguir in the Sahara Desert. Félicette didn't go into orbit, but traveled 160 kilometers into space, where the capsule separated from the rocket and descended by parachute. Throughout the flight, electrodes implanted in Félicette's brain transmitted neurological impulses, and the French Centre d'Enseignement et de Recherches de Médecine Aéronautique (CERMA), which directed the flights, stated afterwards that the cat had made a valuable contribution to research. The capsule and Félicette were safely recovered, but what happened to Félicette after her adventure no one knows, although the epic trip was commemorated on postage stamps from former French colonies some 30 years after the historic voyage.

The most recent suborbital animal flight was carried out by Iran in January 2013, when a monkey was launched in a capsule called Pishgam (Pioneer) to an altitude of 120 kilometers and returned safely (an attempt was made with a monkey in 2011, but the mission failed for reasons unknown). Iran hopes to send a human to space in 2020.

Despite losses, these animals have taught the scientists a tremendous amount and, without animal testing in the early days of the manned space program, the Soviet and American programs could have suffered great losses. These animals performed a service to their respective countries that no human could or would have performed. They gave their lives and/or their service in the name of technological advancement, paving the way for humanity's forays into space, including yours.

Appendix II

This section is included to help you make an assessment of the dangers you may face when you embark upon your trip of a lifetime. It is generally agreed that spaceflight is inherently risky and that adverse physical and psychological effects can be experienced even during successful spaceflights. There are also numerous vehicle and/or system failures that could result in severe injury, dismemberment, or death. The first section summarizes physical hazards and the second section presents potential psychological hazards. Finally, in the third section, there are two tables summarizing these hazards regarding probability of occurrence and severity.

E. Seedhouse, *Suborbital: Industry at the Edge of Space*, Springer Praxis Books,
DOI 10.1007/978-3-319-03485-0, © Springer International Publishing Switzerland 2014

SECTION 1: PHYSICAL HAZARDS

Table A1. Physical Hazards, Causes, and Potential Physical Effects.

Source of hazard	Mission phase/failure mechanism	Potential physical effects
High-decibel noise	Excessive engine noise Inadequate acoustic shielding Explosion on ground	Ear damage Temporary/permanent hearing loss Vestibular effects on balance
High pressure	Breached high-pressure vessel Explosion In-flight aerodynamic pressure	Loss of consciousness Severe ear drum or tissue trauma due to overpressure Concussion Brain damage Death
Low pressure	Explosive decompression Rapid decompression Loss of atmospheric control systems	Trauma due to exposure to vacuum: • Brain injury • Lung injury • Other tissue damage • Death Trauma due to pressure change and trapped gas: • Gastrointestinal pain • Tooth, ear, sinus pain
High G-forces (sustained acceleration)	Acceleration during launch phase, de-acceleration during descent phase	G profile may have adverse physiological on the cardiovascular response of susceptible passengers Cardiovascular Neurovestibular Musculoskeletal
Microgravity	At high altitudes during suborbital flight	Short exposures to microgravity may cause acute physiological responses in: • Cardiovascular system • Respiratory system • Neurological system • Vestibular • Motion sickness • Vision • Musculoskeletal system • Gastrointestinal system
High temperature	In-flight fire/explosion Heat of re-entry/loss of heat dissipation systems	Tissue damage and/or serious burns Death

Source of hazard	Mission phase/failure mechanism	Potential physical effects
Low temperature	Cabin breach	Frostbite/death
Physical impact trauma	Crash/structural failure of spacecraft	Serious injury or death
	Egress from spacecraft	Minor injury
Exposure to toxic chemicals	Release of toxic substance on board	Respiratory/skin damage Death
Electrical shock	Contact with exposed high-voltage source	Severe burns Electrocution/death
Loss of breathable atmosphere/contaminants and particulates	Cabin flooded with non-breathable gases	Asphyxiation/death Brain/organ damage Death

SECTION 2: PSYCHOLOGICAL RESPONSE HAZARDS

Spaceflight participants may experience excessive physiological/psychological response(s) during the spaceflight. Participants prone to responses that could be hazardous to themselves or others should be identified and appropriate measures taken to minimize the risk of the hazards. Table A2 provides some examples of this type of hazard.

Table A2. Summary of Psychological Response Hazards.

Source of physiological/ psychological response/hazard	Potential cause of response	Potential effects of physiological/ psychological response
Claustrophobia	Enclosure in confined space	Excessive agitation Inability to perform required duties
Excitement/Agitation/Fear	Response to unexpected occurrences	Commit irrational and possibly violent, acts
	Response to known risks	Produce anxiety in other passengers
	Mental instability	Incapacitation
Motion sickness	Dynamic motion	Nausea, vomiting Inability to perform required duties/ incapacitation
Vertigo – loss of bearing or balance	Dynamic motion	Nausea, vomiting Inability to perform required duties/ incapacitation
Rapid pulse/Increased blood pressure	Excitement	Cardiac arrhythmia Inability to perform required duties/ incapacitation

SECTION 3: SUMMARY OF POTENTIAL EFFECTS ON PASSENGERS

The business of sending passengers into space is risky. Consider this: approximately 4% of those who have flown in space to date have lost their lives doing so. Four percent! Will commercial space travel be safer? We just don't know, but any vehicle being launched into space may be subject to system or vehicle failure that might result in serious injury or death. Such failures have a variety of potential causes – propulsion system failures, explosion of propellants on the ground or in the vehicle, loss of vehicle control, explosive decompression, and ground impact. The potential hazards of strapping into a spacecraft have been listed in Sections 1 and 2. Tables A3 and A4 rank these effects based on their probability of occurrence and the severity of the resulting consequence. Of course, these are subjective rankings based on typical expectations and will vary from one operator to another.

Table A3.   Potential Hazards – Probability of Occurrence.

Probable/certain
Gastrointestinal issues caused by microgravity
Dysrhythmia (changes in cardiac rate, rhythm) due to acceleration stress
Exposure to actions of other passengers
Somewhat likely
Motion sickness caused by unusual attitude and/or microgravity
Faint feeling caused by acceleration and deceleration
Fatigue caused by low-pressure cabin environment
Panic/fear/fright
Possible
Gravity-induced loss of consciousness caused by acceleration
Moderate injury caused by impacts inside cabin
Vertigo caused by loss of bearing
Significant pulmonary/respiratory effects caused by acceleration in susceptible individuals
Cardiovascular effects caused by acceleration/microgravity
Claustrophobia – hopefully this will have been screened for
Rare – vehicle or safety system failure
Death/severe injury/dismemberment
Asphyxiation caused by loss of cabin atmosphere
Temporary or permanent hearing loss
Burns due to ground accident
Ear drum damage

Table A4. Potential Hazards – Severity of Consequence.

Critical
Death/critical injury/dismemberment caused by vehicle accident
Asphyxiation caused by decompression
Permanent hearing loss
Bone fractures
Loss of consciousness caused by acceleration
Significant
Moderate injury caused by ground failure or impacts within craft
Connective tissue damage caused by acceleration
Ear drum damage
Temporary hearing loss
Vestibular effects – vertigo/balance
Nuisance
Exposure to actions of other passengers
Dysrhythmia in susceptible individuals
Motion sickness
Faint feeling
Headaches
Fatigue
Panic/fear/fright
Claustrophobia
Inability to think rationally
High-altitude sickness

Given the risks of this endeavor, you may be thinking about insurance. This is dealt with in Appendix IV, but be warned that this issue is far from resolved; buying insurance won't cover you to the same extent that regular travel insurance does.

Appendix III

POLAR SUBORBITAL SCIENCE IN THE UPPER MESOSPHERE

The Opportunity

The "Polar Mesospheric Cloud (PMC) Imagery and Tomography Experiment", is a high-latitude campaign selected by the NASA Flight Opportunities Program (Experiment 46-S). It will employ a manned reusable suborbital vehicle that will launch from a high-latitude spaceport (e.g. Alaska or Kiruna, Sweden) during a weeklong deployment scheduled for July 2014 to study the small-scale dynamics of noctilucent clouds. The PoSSUM Project will make full use of the 46-S opportunity by 1) fully utilizing all available payload space to optimize technology maturation and science return for each sortie, 2) combining relevant payloads that would produce alternate missions that could be flown when PMC activity is not observed, and 3) absolving NASA of the opportunity costs that would be incurred by having the spacecraft unutilized while deployed.

The Mission

CAMPAIGN DATES: July 21-28, 2014

LOCATION: Eielson AFB or Kiruna, Sweden

PLATFORM: XCOR Lynx Mark II (proposed)

MISSION PROFILE: When strong cloud formations are observed from the ground or from LiDAR, the spacecraft will be launched to an apogee of 100km, transitioning the PMC layer. The clouds will be under direct illumination from the sun and the attitude of the spacecraft would be oriented north to the presumed region of highest cloud density.

Instrument Suite

Payload Bay B:

PRIMARY MISSION: Operator controlled video camera, still-frame camera, infrared camera, and temperature spectrometer

ALTERNATE MISSION: Operator controlled depolarization LiDAR System, still-frame camera

Cowling Port (CP) Pod:
Meteoric Smoke Experiment (ALL MISSIONS)

Cowling Starboard (CS) Pod:
Mesospheric Winds Experiment (ALL MISSIONS)

Two Ground Observation Stations

E. Seedhouse, *Suborbital: Industry at the Edge of Space*, Springer Praxis Books, DOI 10.1007/978-3-319-03485-0, © Springer International Publishing Switzerland 2014

What are PMCs and why are they important?

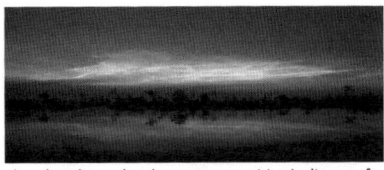

PMCs are the highest clouds in the Earth's atmosphere, 83 km (50 miles) and are observed slightly below the mesopause in the polar summertime. These clouds are of special interest as they are sensitive to both global climate change and to solar/terrestrial influences. The first recorded sightings of PMCs were reported in 1885 and both satellite and ground-based observations over the past four decades have indicated that the presence of these clouds has been increasing in both frequency and brightness. Scientists now realize that these clouds are very sensitive indicators for what is going on in the atmosphere at higher altitudes as small changes in the atmospheric environment can lead to large changes in the properties of these clouds. Further, since these clouds form on condensation nuclei through cold temperatures and the presence of water vapor – and these properties of the mesosphere are tied to carbon dioxide and methane, the anthropogenic causes of climatic change may be directly related to the presence of PMCs.

Technology Benefit:

* Validate a method for repeatable, low cost, in-situ PMC and aeronomy observations.

* Expand baseline PMC observatory to the 'Uppik Observatory', an OCT "Technology Area 8" payload that will serve the ground observation, glaciology, forestry, agriculture, aeronomy, astronomy, and planetary science communities.

Education and Public Outreach:
* Cinematography team established
* Various non-profit EPO organizations engaged

Science Objectives:

* What are the small-scale dynamics of PMCs and what does this tell us about the energy and momentum deposition from the lower atmosphere?

* What is the seasonal variability of PMCs, mesospheric dynamics, and temperatures?

* Are fine structures observed in the OH layer coupled with PMC structures?

* How do PMCs nucleate?

* What is the geometry of PMC particles and how do they stratify?

Mission Management
PI: J. Reimuller, GATS, Inc.

Science Team
G. Thomas, LASP
D. Fritts, GATS, Inc.
G. Baumgarten, Leibniz Institute
Franz-Josef Lübken, Leibniz Institute

Instrument Teams

The PoSSUM team consists of 12 instrument PIs with payloads focused on PMC science.

Partners

XCOR Aerospace   Integrated Spaceflight
LASP   Space Science Institute
Ball Aerospace   GATS, Inc.
Flightline Films   SRI International

Direct questions to Jason Reimuller, GATS, Inc.   (720) 352-3227   jason@integratedspaceflight.com

Appendix IV

INSURANCE

There is very little information about flight insurance for suborbital passengers. Having said that, useful information can be gleaned from the approach to insuring hazardous activities across a range of European and US insurers such as Aviva Irish Life Old Mutual and AXA Legal & General Omaha. While insurers haven't considered the risk aspects of suborbital flight to the point where they have made changes to their process of insuring someone, there is an existing framework for dealing with hazardous activities, under which suborbital spaceflight would fall. But, before we discuss the insurance options, let's consider just how risky the business of launching humans into space is. Perhaps one of the best reference documents describing the risks of manned spaceflight is the Columbia Accident Investigation Board (CAIB) report. Here's an excerpt from that report (my italics):

> "Although humans have been launching orbital vehicles for almost 50 years now – about half the amount of time we have been flying airplanes – contrast the numbers. Since Sputnik, humans have launched just over 4,500 rockets towards orbit (not counting suborbital flights and small sounding rockets). During the first 50 years of aviation, there were over one million aircraft built. Almost all of the rockets were used only once; most of the airplanes were used more often. In the early days as often as not the vehicle exploded on or near the launch pad; that seldom happens any longer. It was not that different from early airplanes, which tended to crash about as often as they flew. Aircraft seldom crash these days, but *rockets still fail between two and five percent of the time*. This is true of just about any launch vehicle [including the shuttle]. It is unlikely that launching a space vehicle will ever be as routine an undertaking as commercial air travel – certainly not in the lifetime of anybody who reads this.
>
> Because of the dangers of ascent and re-entry, because of the hostility of the space environment, and because we are still relative newcomers to this realm, operation

E. Seedhouse, *Suborbital: Industry at the Edge of Space*, Springer Praxis Books,
DOI 10.1007/978-3-319-03485-0, © Springer International Publishing Switzerland 2014

of the shuttle and indeed all human spaceflight must be viewed as a developmental activity. *It is still far from a routine, operational undertaking.* Throughout the Columbia accident investigation, the Board has commented on the widespread but erroneous perception of the space shuttle as somehow comparable to civil or military air transport. They are not comparable; the inherent risks of spaceflight are vastly higher, and our experience level with spaceflight is vastly lower."

One statistic not mentioned in the discussion above is that, when you lay out the raw statistics of human spaceflight, the fatality rate is about 4%. Of course, you can't extrapolate this statistic to commercial suborbital spaceflight because this is not an actuarial business; it is unlikely you have a 4% chance of dying on your suborbital ride to space. But, even if you did, how does that number relate to other hazardous activities? After all, plenty of other ventures exist with similar fatality rates; in recent years, about 4% of the people climbing Mt. Everest have died (in the late 1970s, that number was higher). For the mountain-climbing community, a fatality rate of 4% has not deterred people from taking the risks – in fact, the numbers of climbers heading to Everest increases every year. In short, it's impossible to completely eliminate the risk of human space travel. In fact, as commercial manned spaceflight becomes a reality in the very near future, the existence of that element of risk may even prove to be a key selling point: those who have climbed Everest perhaps? In fact, it is highly likely that many of the early passengers on suborbital spacecraft will be thrill-seekers who willingly accept a chance of death to fly faster and higher than ever before. And it will be these same people who will provide the revenue that will allow companies to build new generations of safer spacecraft. So, how do they go about getting insurance?

APPLICATION PROCESS

If you want to take out insurance for your flight, you will probably have to complete a hazardous activities form similar to the one below.

Aviva Term Application Form

Which of the following activities do you take part in:

Motor sport	☐
Climbing	☐
Diving	☐
Caving/potholing	☐
Extreme sports	☐
Equestrian sports	☐
Flying (other than as aircrew of fare-paying passenger)	☐

Source: AVIVA, 2012.

Other insurers have different ways of capturing participation in hazardous activities. For example, some ask an open question: Do you take part in any hazardous activity? e.g.

private aviation, diving, yachting, or sailing, mountaineering, or rock climbing, motor sports, caving, or potholing, parachuting, hang gliding. Or: Have you in the last 5 years or do you intend to: Take part in any sport or pastime which involves any additional risk of accident? e.g. aviation, motorsports, climbing, diving, horse riding, martial arts or winter sports (Prudential, 2012). Needless to say, if you intended to fly a suborbital flight, you would have to disclose that! Chances are that, after you've disclosed the fact you're planning on flying a suborbital flight, the insurer will want you to complete their aviation questionnaire, which will probably cover you for the flight. Probably.

Once your application is submitted, the underwriting process begins to assess the risk of your flight – a process that may require you to answer more questions about your suborbital ride. Ultimately, the decision to insure you will be based on an assessment of risk. If the risk is considered too high, insurance will be declined. Also, if a clear understanding of the risk cannot be obtained, cover can be postponed. Alternatively, the insurance company may ask you to pay an additional premium to cover the risks. The ratings that the insurance company calculates may be temporary or permanent, and may be expressed as a flat addition or as a percentage of additional mortality, which is the normal approach to covering aviation risks and may well be the favoured approach by insurance companies to insuring suborbital flights. If insurers consider the death risk from suborbital flight mirrors the spaceflight experience to date (4% mortality rate), they may impose ratings several times a typical private aviation rating, which could prove expensive.

One of the few companies that provide insurance to suborbital astronauts is Allianz Global Assistance. In November 2011, Allianz Global Assistance and the International Space Transport Association (ISTA) announced a partnership to provide a travel insurance package for suborbital passengers. Although there is little information about the details of the policy, at the 2012 Royal Aeronautical Society's Space Tourism Conference, the following information was provided:

- Although Allianz will consider space tourism in the same general risk category as high-risk sports such as mountain climbing, they are targeting space tourism as a much higher volume activity than adventure sports.
- It will be not be a single product, but a base cover with up to 20 selectable options.
- Allianz see their distribution channel as being through the operators, rather than through insurance intermediaries or direct to the public, enabling them to build a specific product for each operator. For example, there could be different options to cover the training activities and/or the actual trip.
- Premium rating factors will not include age (*except perhaps for very high ages*), but will include nationality, due to variations in national insurance obligations.
- A typical additional premium for high-risk categories is 3–5%. However, the premium for this cover is likely to be moderate. This is because a significant component of the overall premium is cancellation risk. Most of the operators have a cancellation clause imposed by FAA, for example Virgin Galactic will return 95% of the ticket price on cancellation. Because of the extreme high-end nature of the experience and the scarcity of supply, Allianz feel that cancellations will be rare. At the time of the conference, the premium was estimated to be EUR500–700 per trip (which is about $650 to $900).

Index

A

Acceleration, 6, 23–29, 46, 47, 60, 62–68, 75, 90,
    136–138, 149, 151, 170, 172, 173
Adams, Mike, 68, 69
AGSM. *See* Anti-G straining maneuver (AGSM)
AGSOL. *See* Ashton Graybiel Spatial Orientation
    Laboratory (AGSOL)
Aldrin, Buzz, 16, 44, 87, 89, 117, 118, 162
Allen, Paul, 18, 88
Almost Loss of Consciousness (A-LOC), 63
Alsbury, Mike, 85
Anti-G Straining Maneuver (AGSM), 26, 28, 65,
    78, 82
Apollo 13, 51–53, 155
Armadillo Aerospace, 1, 34, 97, 102–106, 117
Armstrong Line, 71
Armstrong, Neil, 2, 4, 11, 68, 71
Ashton Graybiel Spatial Orientation Laboratory
    (AGSOL), 83
Astrochimps
    able, 5
    astronauts for hire (A4H), 41, 79–81
    Baker, 5
    Gordo, 5
    Ham, 6, 7, 166
Astronauts for Hire (A4H), 31, 41, 46, 52, 70,
    78–81, 114, 132
AXE, 44, 161, 162
    AASA, 161

B

Bail-out, 28–29, 78
Baumgartner, Felix, 29
Bezos. Jeff, 2, 98, 99, 101, 102
Bigelow Aerospace, 52, 118

Binnie, Brian, 17, 18
Black Brant, 48, 50
Blue Origin
    Goddard, 99, 101
    New shepard, 34, 38, 98, 99, 101,
        152, 153
Borozdina-Birch, Lina, 45
Branson, Richard, 18, 44, 85, 87–89, 92, 116, 118,
    127, 156

C

Canadian Space Agency (CSA), 36
Caribbean Spaceport, 117, 118
Carmack, John, 102
Commercial Crew Development (CCDev), 99
Commercial Crew Program (CPP), 99
Commercial Space Launch Amendments Act
    (CSLAA), 19
Copenhagen Suborbitals
    Horizontal Assembly Building (HAB),
        107, 108
    Madsen, Peter, 109, 111
    Tycho brahe II, 109
    von Bengtson, Christian, 107, 109, 111
CPP. *See* Commercial Crew Program (CPP)
CSA. *See* Canadian Space Agency (CSA)
CSLAA. *See* Commercial Space Launch
    Amendments Act (CSLAA)
CubeSat, 39, 53, 54, 145, 149, 150

D

DCS. *See* Decompression sickness (DCS)
DC-X, 99, 101

E. Seedhouse, *Suborbital: Industry at the Edge of Space*, Springer Praxis Books,
DOI 10.1007/978-3-319-03485-0, © Springer International Publishing Switzerland 2014

Decompression, 20–23, 30, 71, 72, 136, 150, 151, 170, 172, 173
Decompression sickness (DCS), 22, 23, 74
Defence Advanced Research Projects Agency, 101
DiZio, Paul, 69
Durda, Dan, 77, 126, 129

E
Effective performance time (EPT), 21, 22
Emergency egress, 28, 30, 59, 70, 75, 76, 78, 80, 82, 110, 136, 145
Engle, Joe, 68
Environmental Tectonics Corporation (ETC), 77
European Space Agency (ESA), 36, 61, 82, 121
Evolved Expendable Launch Vehicle (EELVs), 48, 54
Expedite the Processing of Experiments to Space Station (EXPRESS), 37

F
Federal Aviation Administration (FAA), 15, 18, 19, 32, 33, 54, 59, 67, 74, 75, 77, 92, 102, 104, 155, 158, 179
Fédération Aéronautique Internationale, 12, 38, 59, 68
Final Frontier Design (FFD), 21, 22, 72
Flight Opportunities Program, 49, 58, 106, 125, 158, 160, 161

G
Galactic cosmic radiation (GCR), 31, 72, 73
Garver, Lori, 58, 158
Gradual onset rate, 27, 64, 65, 67
Gravity-Induced Loss of Consciousness (G-LOC), 23–27, 63, 64, 172
Grissom, Gus, 9–11

H
Howard, Mindy, 81

I
Innerspace training, 81–83
International Committee on Radiological Protection (ICRP), 73
International Space Station (ISS), 36, 39, 46–49, 52, 73, 99, 127, 141, 157, 160

K
Kaplan, Janna, 69
Kittinger, Joe, 29

L
Lackner, James, 69
Lovelock, James, 45
Lynx
    Mark I, 1, 34, 49, 94, 95, 148–150
    Mark II, 1, 34, 94–96, 133, 148–150
    Mark III, 34, 54, 94–96, 148, 149

M
Masten
    Lunar Lander Challenge, 104, 106
    Scimitar, 107
    Xaero, 34, 106, 107
    Xeus, 106
    Xoie, 106
    Xombie, 104–107
Melvill, Mike, 2, 18, 88, 90
Mojave Spaceport, 122–124

N
NanoRacks, 37, 39, 127, 141
National Aeronautics and Space Administration (NASA), 2, 4–8, 10–14, 18, 22, 24, 31, 36, 47, 49–52, 53, 54, 58, 59, 61–65, 68, 69, 73, 80, 87, 91, 99–102, 106, 107, 117–119, 125–129, 132, 138, 145, 152, 158, 160–162
National Aerospace Training and Research (NASTAR), 43, 59, 77, 78, 80
National Space Biomedical Research Institute (NSBRI), 47
Nelson, Andrew, 49, 92, 94, 118, 151
New Mexico Spaceport Authority (NMSA), 116, 117
Next Generation Suborbital Researcher's Conference (NSRC), 1, 2, 4, 11, 37, 44, 58, 126, 127, 150, 152, 158
Nield, George, 32
NMSA. *See* New Mexico Spaceport Authority (NMSA)
NSBRI. *See* National Space Biomedical Research Institute (NSBRI)
NSRC. *See* Next Generation Suborbital Researcher's Conference (NSRC)

O

Olkin, Cathy, 126, 162

P

Payload user's guide, 106, 141, 143, 151
Polar Suborbital Science in the Upper Mesosphere
    (PoSSUM), 127, 131–133
Principal, Victoria, 45, 92
Project Mercury, 5–11

R

Radiation, 5, 30–32, 48, 60, 72–75, 150, 160
Rapid onset rate (ROR), 64, 65
Reaction Rocket Engines, 42, 57
Reimuller, Jason, 79, 93, 131, 133
Research Education Mission (REM), 53, 77
Research Opportunities in Space and Earth
    Sciences (ROSES), 161
Roski, Edward, 45
Rutan, Burt, 16, 18, 87–91, 113, 155–157

S

SARG. *See* Suborbital Applications Research
    Group (SARG)
SAS. *See* Space Adaptation Syndrome (SAS)
Scaled Composites, 2, 12, 14, 16, 17, 19, 89, 91,
    94, 101, 123, 156, 157
Shatner, William, 16, 89, 156
Singer, Bryan, 44
SIRIUS Astronaut training, 69, 83
SMS. *See* Space Motion Sickness (SMS)
Solar Cosmic Radiation (SCR), 31, 73
Southwest Research Institute (SwRI), 4, 77,
    125–127, 129, 158, 162
SPA. *See* Suborbital payload agent (SPA)
Space Adaptation Syndrome (SAS), 137
Space Motion Sickness (SMS), 46, 47,
    69–71, 83, 137
Spaceport America, 86, 87, 90, 116, 117, 123,
    124, 142, 145, 150, 151, 156
Spaceport Sweden
    Esrange, 119–121
    IceHotel, 119, 121
    Kiruna, 118–121
SpaceShipOne (SS1), 1, 11, 13–19, 45,
    59, 60, 74, 87–89, 90, 101, 113,
    123, 157
SpaceShipTwo (SS2), 1, 4, 12, 33, 34, 40,
    46, 57, 60, 67, 78, 85–92, 94, 95, 101,

    123, 125–127, 137, 141–148,
    156–158, 162
sRLVs. *See* Suborbital reusable launch vehicle
    (sRLVs)
SS1. *See* SpaceShipOne (SS1)
SS2. *See* SpaceShipTwo (SS2)
Stand test, 91, 109, 130, 131
Stern, Alan, 4, 77, 125, 126, 129, 158, 162
Stucky, Mark, 85
Suborbital Applications Research Group (SARG),
    126
Suborbital payload agent (SPA), 81, 82, 153, 154
Suborbital reusable launch vehicle (sRLVs),
    33–36, 38–42, 44, 46–49, 51, 53, 55–58,
    106, 115, 161
Suborbital training, 59, 67, 70, 81, 82, 138
Sundlisaeter, Tale, 162, 163
SwRI. *See* Southwest Research Institute (SwRI)

T

Tauri Group, 33–38, 40, 42, 45, 48, 49, 53, 54, 58
Time of useful consciousness (TUC), 20

U

Uninhabited aerial vehicle, 55, 56
United States Air Force (USAF), 12, 28, 59,
    68, 71
United States Munitions List (USML), 122, 158
United States Navy (USN), 6, 12, 65
Universities Space Research Association (USRA),
    126, 159
USML. *See* United States Munitions List (USML)
USN. *See* United States Navy (USN)
USRA. *See* Universities Space Research
    Association (USRA)

V

Vertical take-off/vertical landing (VTVL), 30,
    100, 104, 107, 152
Virgin Galactic, 1, 3, 4, 13, 34, 40, 44, 49, 54, 60,
    76–78, 85–92, 94, 97, 103, 111, 116–118,
    121, 123, 125–128, 141–149, 153,
    155–157, 161, 162, 179

W

Walker, Joe, 16, 68
WhiteKnight, 12, 15, 16, 88
WhiteKnight 2 (WK2), 54, 60, 85, 142, 157

Whitesides, George, 40
Wright, Ed
  Citizens in Space, 158, 159
  Space Hacker Workshop, 159

X
X-15, 2, 4, 11–16, 18, 59, 67–69, 87, 88, 156

XCOR Aerospace, 1, 34, 49, 76, 92, 118, 123,
  129, 162
X-Prize, 1, 16–18, 86–88, 102, 113

Z
Zero-G Corporation, 52

Printed by Printforce, the Netherlands